AF722588

PRAISE FOR WINNOVEX™

"Adifon is a master of the craft of engineering, with a clear vision of the big picture, and an impressive command of the details. WinnoveX represents the synthesis of more than 35 years of hard-won experience at global companies; 35 years of delivering innovative products that have delighted customers and won profits. Adifon offers us a blueprint for his success from which even seasoned engineering leaders can learn much."

— Michael Wynblatt, Vice President and Chief Technoloy Officer, Donaldson Company, Inc.

"We all see things from our own experience. Leandre's experience covers the breadth of the product life cycle across multiple product domains and around the world. In addition, he has the knack for observing a situation and thinking about how to make it better. These capabilities and experiences are shared with you in WinnoveX: Win Your Way to Innovation Excellence."

— Ginny Lentz, Systems Engineer

"Whether you are an executive needing to enhance innovation capabilities or starting your journey as a new product development leader, or want to further your understanding of innovation systems, WinnoveX: Win Your Way to Innovation Excellence can help. In this book, Leandre presents a complete and easy-to-understand set of key principles to define and execute, customer-centered innovation."

— Manolo Caballer, VP Innovation & NPD Kitchen & Bath, Kohler Co.

"Rooted in systems thinking, WinnoveX provides a structured and holistic approach for innovation planning and execution. I highly recommend it for hard tech innovators who develop complex products with long life cycles."

— Dr. Vijay Jayachandran, CEO at ACW Analytics and Author of *Inpowering Leadership*

"A seminal book about innovation, product development and business. Adifon will transform your mind and help you see the world in a new way. His innovation methodology can transform any industry. WinnoveX gives you a clear path to build and scale billion-dollar products, startups or enterprises. Whether you're an entrepreneur, product manager, innovator or Fortune 1000 CEO, this is a must-read innovation bible."

— Damir Perge, Co-Founder and CEO of both FluidLytix & Entrepreneurdex, Author of *Futbolpreneur* and *Entrepreneur Myths: The Startup Reality*

WinnoveX™

Win Your Way to Innovation Excellence

LEANDRE ADIFON

Pyramid Base Technologies, LLC
Charlotte North Carolina

2020

WinnoveX: Win Your Way to Innovation Excellence.

Pyramid Base Technologies, LLC
Publishers since 2020
www.pyramidbasetechnologies.com

Copyright ©2020 by Leandre Adifon

All rights reserved. No part of this book may be reproduced or transmitted in any form or by any means, electronic or mechanical, including photocopying, recording, or by informational storage or retrieval systems, without the written permission of the publisher.

For information on discounts for bulk purchases, contact Pyramid Base Technologies, LLC at:

info@pybtech.com

Printed in the United States of America

Library of Congress Cataloging-in-Publication Data

Adifon, Leandre
WinnoveX™: Win Your Way to Innovation Excellence /
Leandre Adifon – 1st ed.

Includes bibliographical references and index.

ISBN 978-1-7361617-0-8

1. Innovation 2. Business 3. Engineering 4. Product Design 5. Product Management

First published in print format in 2020
Pyramid Base Technologies Edition – 2020
Cover design by MadChita Ventures, LLC
Cover photography by MadChita Ventures, LLC

This book would never have been possible without my wife Josiane's support and encouragement. Her patience during all the long hours I was home without being present was an invaluable asset.

To Joel, my son, who brainstormed the book's title with me and helped me pick the final one among tens of possibilities.

To my parents, who made great investments in me but did not live long enough to appreciate the results of their sacrifices.

LEANDRE ADIFON

To all who believe that the discoveries and inventions not yet made will be at some point in time and will continue to improve people's lives and preserve planet Earth.

LEANDRE ADIFON

Acknowledgment

To my bosses and mentors across several companies and continents who have been patient and trustful even when my unconventional manners were not always understood nor welcome.

To all the reviewers of the manuscript who provided valuable feedback to improve my study. Pat Hale, Daryl Marvin, Bruce Lerner, John Bergh, Paul Romanello, Vlad Uherek, Evan Gerber, your time reading the manuscript and challenging some of the concepts allowed me to refine my thoughts.

A special shout goes to Jana Arnold and Uday Deshpande, who took the time and had the patience to edit the book, making sure other languages I speak do not get in the way of the concepts expressed.

This book is the synthesis of the contribution and personal investments of many people. It is also the result of more than 30 years of observations and practice, first as a consultant and then an employee in large multinational companies.

During my career in the private sector, I have been a constant learner from my bosses, peers, teammates, and other people from the companies I worked for because I am convinced that we can learn from everyone.

Contents

List of Illustrations

Foreword

The speed of life in modern society has pushed quite a few people, unbeknown to them, away from observation and contemplation. How many times do you stop everything you are doing with your busy schedule to look at life around you? If you have stopped doing it or if you are not used to contemplation, please get back to it or start doing it. You will be surprised to discover how many wonders surround us. Nature offers trillions, quadrillions, of harmoniously crafted examples of elements that form a whole. Like in a perfect orchestra, each part plays a role, or in other words, "performs a function or set of functions" to enable the whole (the system) to fulfill its mission.

Whether it is a natural being (a biological system) or engineered equipment (an artificial system), the harmony can be striking and, once observed, can captivate our full attention for a moment before we consider the next one. I continue to enjoy those moments when I can appreciate the beauty of a flower, a butterfly, or when I stop in front of a machine, be it a bicycle or as complex as an airplane. As an engineer and passionate about product design, I have often borrowed from nature, its beauty of well-architected beings to design new products. Such a practice, however, is not new. Over the centuries, product designers have borrowed great concepts from nature to generate winning products.

Before getting into the core of the book, I wanted to share a couple of observations. They would help in the next chapters since I will be highlighting statements and conclusions drawn from similar explorations.

1. A Plant

My first sharing is about a plant. Years ago, a small flower plant my wife received in a transparent vase captivated my attention. She placed it on the kitchen window to catch the beautiful sun's rays reaching our house every morning. I could see part of its roots in the dirt. Its tiny trunk grew out of the vase as it supported a few branches that bore the leaves and a few flowers. It was a full system. It interacted with its environment by receiving inputs and providing some outputs. Like any other plant, through photosynthesis, it provided oxygen to its surroundings. Its flowers shared their beauty and perfume with those who, like me, had the curiosity to observe them. I have contemplated that simple system for many weeks.

Fig. 0.1 A Plant

The roots' function was to absorb the water and the other nutrients necessary for the plant's growth. They would then send them to the trunk, the branches, the leaves, and the flowers. The same roots anchored to the dirt provide stability to the plant to prevent it from toppling over. Above the soil, the trunk represented the main structure of the plant. It supported the branches and acted as a pipe for the sap to reach the leaves and flowers. All the structural elements had the function to bear the leaves and drive nutrients to them. I was fascinated by the leaves. They were real machines capable of complex processes. Without getting into biochemistry to explain the complex process by which carbon dioxide, water, and certain inorganic salts are converted into carbohydrates using chlorophyll and energy from the sun, I would simply state that they enabled photosynthesis. The last elements of the plant system were the flowers. In general, they would allow reproduction and generate fruits. They also shared their beautiful colors and perfume for the eyes and nose of the admirer.

One Saturday afternoon, as I was observing the plant, I could not help but draw Figure 0.1.

Here is one more observation from the repertoire of my mind travel collections.

2. A Butterfly

The second system from my observations is a butterfly. I have spent countless minutes looking at butterflies in our backyard. We had planted a "butterfly bush" in our garden in Connecticut. As the tree grew tall, I could observe the beautiful creatures from the deck at the back of the house. Then I went to learn about them.

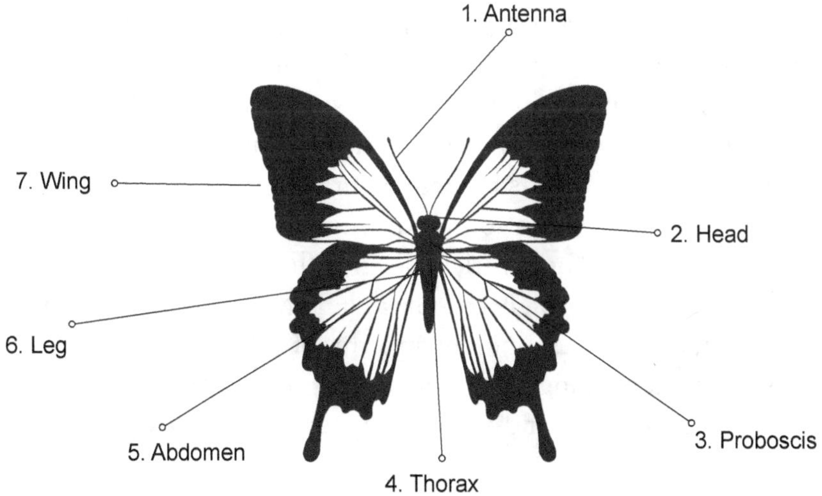

Fig. 0.2 A Butterfly

I found one of their descriptions online that perfectly matched my observations. It was from an article entitled "What is a butterfly?" I found it so fascinating that I quoted it here:

"Butterflies are beautiful flying insects with large scaly wings. Like most insects, they have six jointed legs, three body parts, a pair of antennae, compound eyes, and an exoskeleton. The three body parts are the head, thorax (the chest), and abdomen (the tail end).

Tiny sensory hairs cover the butterfly's body. Its four wings and six legs are attached to the thorax, which contains the muscles that make the legs and wings move. An essential ability of those insects is to fly."[1]

The description above may leave the reader wondering why all these details, even though this book is not a research report on butterflies.Well, I want to achieve two goals by sharing two examples from my collection of simple observations.

First, it is to invite all the readers, whether they are designers, engineers, other professionals, or executives involved in products or organization development, to take the time to contemplate and enjoy the wonders around them. It will bring joy and peace to your innovation mind because there is so much simplicity of design incorporated in the complexity of our life.

Second, a more professional reason is to introduce the notion of "function" as what an entity or any of its constituents does. The next chapters will get into more details about functions, especially how to define them and their role in systems design. But I can highlight right away that thinking in terms of "functions" is not common, and school systems across the world did not attempt to incorporate this type of methods in their classes either. There was a lack of teaching about "functional thinking" in school programs. Even in their profession, most engineers think more "physical" than "functional." Over the decades, engineering primarily focused on mechanical devices, so thinking in terms of physical components became natural. In today's highly competitive environment, electronics, controls, and software need to be designed from a functions perspective as they work to develop their products functional behavior. As the products increasingly become a combination, i.e., electro-mechanical or focused on mechatronics, thinking in terms of functions is becoming crucial.

Functions are a vital ingredient of systems design and innovation. Therefore, we will explore them and know them better in the next chapters.

---

1 What is a butterfly? www.enchantedlearning.com/subjects/butterfly/allabout/index.shtml

1

Introduction

1.1 NEEDS AND EXECUTION

The 21st century is unveiling a trend where there is a proliferation and acceleration of products being released into the marketplace at an unprecedented rate. Due to fierce competition, conventional product development, which requires a market study to understand and interpret customers' needs and design products to meet them, is not sufficient anymore to line up customers wanting to buy new products. Two additional drivers seem to be generating success in the marketplace today:

The first one is about the ability to identify the "conscious" and "unconscious" unmet needs of the targeted customers. As strange as it may appear, people can be unconscious of some needs until someone unveils them through new systems or solutions. Those potential customers would not naturally come forward to express such needs. Until people saw a ballpoint pen for the first time in 1938, they never indicated the need to have one. Everyone was happy using fountain pens or dip pens while learning to cope with their messiness. Then, Laszlo Biro, a journalist from Hungary, introduced the ballpoint pen that enabled the ink to dry faster without leaking on the paper. This innovation became so successful that its design is still in use to this day.

Considering another example, if people were asked in the 19th century what their transportation needs were, their primary request would have been to get faster horses that produced less manure. In fact, at that time, most of the largest cities in the Western world were facing a manure crisis due to the increasing transportation need and subsequent higher population of horses.

Nobody would have expressed the need to have a car. But at the beginning of the 20th century when Henry Ford introduced his Model T, the necessity of having a car became quickly evident in the United States. The affordable vehicle satisfied the conscious unmet need to replace animal-drawn carriages. It also triggered a few unconscious unmet needs with features such as speed, ride comfort, and others.

The second success driver is about execution or delivery. A flawless execution brings a product's value to life. It highlights the attention and care to detail dedicated to its fulfillment. It also drives on-time delivery. In every industry, premier companies distinguish themselves from their competitors by leveraging execution as one of their most potent strategic and operational weapons. Overall, an excellent performance brightens the customer experience.

To fully identify the unmet needs, companies must learn to anticipate their customers' experience with their product or service. It is as if they could observe them living in the new world or ecosystem their product will generate. Then, a good execution by the design team[1] is crucial to highlight the value brought by the satisfaction of the identified needs.

Some companies like Apple Inc. are re-inventing devices such as watches that have existed for a long time by leveraging both the "unconscious" unmet needs and excellent execution. The notion of unveiling or awakening an unconscious need should not surprise anyone. However, it has triggered some discussions every time I raised it. Therefore, I must clarify it due to its considerable implications. The mere notion that there could be unconscious needs pushes market surveyors to look for them by supplementing the traditional Voice of the Customer (VOC) approaches with scenarios likely to generate new habits that would come alive with the use of the product or service. Standard VOC mostly captures the "conscious" needs, even when their statement is not very clear. But a product is likely to lead the market if what it offers not only satisfies the customers' conscious unmet needs, but it also triggers some unconscious unmet needs. VOC research evolution is making fair use of observations beyond the conventional "questions and answers" sessions. Observing the users in their natural environment as they interact with existing products or concept products to identify improvement

1 In the context of the book, the design team is assumed to include the functions of engineering, manufacturing, purchasing, installation, service and human resources.

opportunities or potential next-generation designs has been used for decades. These methods have their merit and help identify some types of customers' needs. But they are subjective due to their link to human behavior and are less deterministic in leading to the identification of the unconscious unmet needs. **The approach that is taken in this book to discover unconscious unmet needs relies on understanding the functions a product performs or could perform in the future. The next chapters cover various techniques for the identification of what a system does or could do in the future.**

As an inventor and engineer, I conducted a lot of informal surveys of customers of various products around the world in the last couple of decades. I could observe that the satisfaction of an unmet need that people did not know they had often left them wondering how they lived their lives without such a product or service. Designing or improving any system starts from defining or understanding its mission, the outcomes it must drive, or what it will do, all of which are expressed through its functions as they discover the unmet needs it is conceived to satisfy.

1.2 THE SYSTEMATIC INNOVATION OPPORTUNITY

In the course of history, the world continuously evolved, and human conditions improved through new solutions. Several dictionaries define innovation as the act of introducing something new and needed. When we innovate, we present something that did not exist before its release. We unveil solutions that open new possibilities that were neither contemplated nor imagined. In their quest to provide new systems with better attributes and outcomes than the ones in use, innovators must act as if they were knocking down the walls that divide their current world from the new one, they are creating. Unfortunately, a commonly shared view across cultures and companies is that "only" visionaries can innovate. But even though visionaries do innovate, the assumption is not accurate. Numerous studies and methods attempted to make innovation more systematic and less dependent solely on the genius of a few individuals. Most of the suggested theories, however, are so intricate they need highly educated experts to effectively execute them. Therefore, the need for a universal and straightforward approach to make innovation systematic is still open. The fascinating history of sciences has shown that practice often precedes theory. The ways of achieving a solution would exist for decades or even centuries before their scientific formulation could be achieved and widely shared. The last two centuries have seen significant progress in creating unifying theories that have started to merge various domains of knowledge. Still, there continues to be the need to descent deeper to the heart of the unifying principles of the innovation of entities such as machines and human organizations. Such a dive would frame theories derived from decades of best practices to consistently conceive winning solutions.

I had the opportunity to author more than 30 patents of successful products while working for large corporations. I also facilitated countless sessions across geographies that led to several innovations. Such experience led me to realize that a common thread links people's needs, functions, technology, and execution to drive the creation of any system such as a product, a process, a service, or even an organization. The same relationship between the elements that make the thread mentioned above paves the way to sustained innovation.

1.3 INNOVATIVE PRODUCTS OFFERING

One of the main goals of companies involved in design, manufacturing, and sales is to offer "things" in terms of devices, machines, services, processes, or even organizations that meet the customers' or users' needs. To grow the business, it became crucial to produce them at the lowest possible cost. So, the ability to offer a product with high perceived value at a relatively low cost has been one of the most critical competencies companies pursued for a long time. Market leaders evolved and fine-tuned their competitive strategies over time leveraging new processes and technologies. They made trade-offs and progress along the journey characterized with the three distinct periods represented below in Fig. 1.1.

- Limited and individually customized offerings period
- Mass-produced offerings period
- Mass customized offerings period

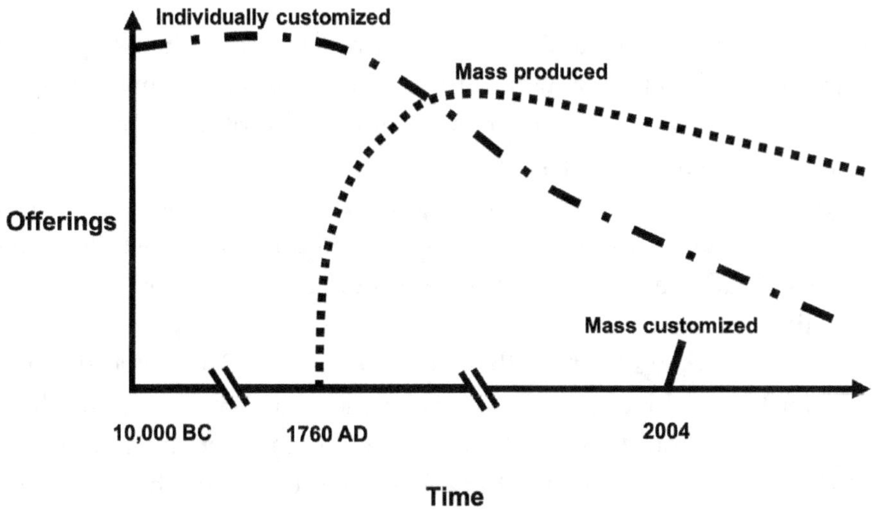

Fig. 1.1 Products offerings over time

There are lessons to draw from each of the periods that will help us on our innovation journey.

1.3.1 Individually Customized Products

Craftsmanship offers a glimpse into a world where every product was different from the next one made, though performing the same function. Whether it was a painting, a basket, a tool, or art objects, there were differences from one product to another. They could be close, but they were not the same. One could argue that no products are the same, even in the modern world, due to features and tolerances on dimensions. For crafts, however, the designer and producer believed they were the same. The design and manufacturing intent were to make them the same. Still, several variables drove their differences. Each artisan's skills, personality, taste, and mood, etc., and from the customers' side, their needs and expectations are just a few. As a result, each offering was different.

Growing up in my native Benin in West Africa, I have observed countless artisans doing their job. Most of the time, the design was all in their head. There were no blueprints, no drawings, or sketches. They would exhibit multiple articles and claim that some were the same. In one of the regions of Benin, some families were, and still are, specialized in specific crafts. Some were tailors; others were blacksmiths, woodworkers, copper workers, leather workers, etc. Several families had carried for centuries the tradition of their specialties assigned to them from the time when kings were ruling the regions. Later, living in Europe, I have visited ceramics artisans in France and Italy with the same observations. Beautiful ceramic articles were made and sold in various stores. No two were identical. The themes, although close, always presented some differences. Variability was part of the products' value. Customers appreciated and purchased them for their uniqueness. Craftsmanship has served well the business to consumer (B2C) relationship.

I was impressed by a blacksmith in my small town of Benin specialized in mousetraps. He would forge and exhibit every kind of a trap. Some resellers from other towns would visit and place large orders. Then, he and his son would work long hours for days to deliver what they thought were traps with the same size, features, and price.

Customizing and delivering to a buyer what he or she specified did not stop in the past. It is still going on today in the industrialized world and is well known as products "Engineered to Order" or ETO. We will revisit the terminology and practice in Chapter 7.

1.3.2 Mass-Produced Products

While the practice of individually customized products has been the norm for centuries, a significant shift started in 1760 when the agrarian and handicraft economy saw the introduction of production machines. The considerable change, known as the industrial revolution, began in Britain and spread over time across Europe and to other parts of the world (as highlighted by the Encyclopedia Britannica) driven by the convergence of several technologies. The use of iron and steel with coal as a source of energy and the invention of the steam engine have opened new possibilities. The creation of the spinning jenny propelled the textile industry with a substantial socio-economic impact. The process of combining the division of tasks and the standardization of parts has led to mass production. Factories could produce large quantities of various components with little variations from batch to batch. Over time, processes and instructions minimized operator differences. The parts were combined to configure different products to meet market demand while keeping the cost low. Some businesses excelled as part or component suppliers to other companies in a Business to Business (B2B) relationship. Mass production is still in force today, and one of its merits is to drive affordability, especially for mid-class customers. Most companies that produce consumer as well as industrial products, see it as the only way to stay relevant in a very competitive world. To manage variation and boost productivity, large corporations designed or adopted excellent production and quality systems. Toyota Production System, for example, has proven that waste across the value chain, due in part to inconsistency and overburden, can be controlled to significantly improve quality and drive high productivity by producing parts and products just-in-time while keeping the work-in-progress inventory to a minimum.

1.3.3 Mass Customization

As mass-production continues to power value generation in the consumer and the industrial world, a trend is emerging with characteristics of both individually customized and mass-produced products. An effort is underway in companies around the world to get as close as possible to giving each customer what he or she wants but still keeping production cost affordable in the current markets. Industrial leaders I had the opportunity to talk to unanimously share the same conviction. Being able to combine the best of both worlds (individually customized products and mass production) will undoubtedly give any company a competitive edge. So, the critical question to answer is: Is it possible to mass-produce an individually customized product? Fortunately, the answer is "Yes," and it is one of the benefits of the journey to Winning Innovation Excellence or *WinnoveX,*™ which is the subject of the book. *WinnoveX*'s adoption can help companies achieve mass customization quicker and more efficiently. A simple example will shed some light on the notion of mass customization.

Every time I go to a restaurant with a salad bar, I cannot stop glancing at the plates of all the diners to see what they have chosen to eat. It is nearly impossible to see two people with the same selection. I am always fascinated as I observe the continuously variable dishes that seem to be infinite choices generated from a limited number of items. Math combinations show that if twenty different items are in the salad bar, and each customer must pick only four in an unordered manner, there would be 4,845 different combinations of servings. To recall the math and to generalize the notion, C being the combinations of n items and r the number of items in each selection, we use the following formula the formula for C (n, r):

C(n,r)=n!/r!(n-r)! ;

so, with the example of the choices mentioned above, we would have:

C(20,4)=20x19x18x17/4x3x2 = 4,845

If instead there were 60 items on the bar and each selection would contain any 8, the number of choices would be greater than 2.5 billion.

The example shows how many distinct solutions can be available from a limited number of defined elements. In the restaurant's case, every customer is happy with his selection as it reflects his preference. Customers decide what they want and value their choices. A B2C transaction is successful when the customer is happy to get his or her unique solution while the supplier delivers it at the minimum cost to maximize profit. Everyone wins.

The key to cost control is to design of a limited set of essential elements with combinations to provide solutions that meet the individual customer or similar customer groups' demands.

In the next chapters, we will explore the elements and steps that lead to mass customization for both consumer and industrial products with a significant return on investment or sales. For any system designed, produced, and offered, the critical starting point is an in-depth knowledge of its function or functions through analysis, as shared in Chapter 2.

1.4 A COUPLE OF NEW CONCEPTS

In the previous sections, reference was made several times to business to consumer and business to business relationships defined as B2C and B2B, respectively. When observed through the periods described in Fig. 1.1, the two types of relationships and their resulting transactions have evolved along two paths.

B2B was either the offering or purchase of products, such as means of production that make the customer's business possible or the purchase of parts integrated into a full product. In this case, the measure of value is the "productivity" driven by the acquisition. Productivity could be assessed here through the improvement of the production process or, in general, the achievement of more value with the same resources. It can also be viewed as having the same value with fewer resources or even more value with fewer resources. In a B2B relationship customers mostly know what they want in full detail. They provide their requirements in the form of written specifications.

In the case of B2C, the opportunities are less clear. The customers often do not know what they want, and even when they think they do, they may change their mind when a solution triggers their unconscious needs. Companies designing, manufacturing, and offering the goods to the consumers bear the burden of anticipating their needs. In this case, "Productivity" is also a significant measure for innovation; it can measure how much time or money the consumer saves by using the product or service, which allows him to earn more available time for other life activities. Therefore, we can say that it is also the opportunity to achieve "more with less."

So, while one type of innovation happens within a well-lit environment with requirements and expectations clearly defined, the second type is somewhat fuzzy with a lot of uncertainties. We will describe them as "Demand-Driven Innovation (DDI)" and "Opportunity Driven Innovation (ODI)," as summarized in Fig. 1.2 below:

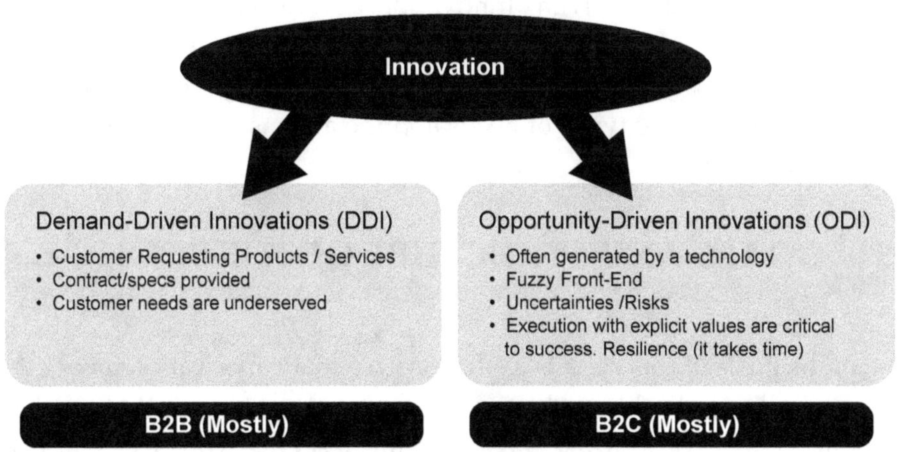

Fig. 1.2 Types of Innovation

Being able to distinguish between the two types of innovation is crucial to success. Making investments and setting expectations depend on the types considered. Unfortunately, it often happens that companies underestimate the uncertainties and risks relative to ODI or simply ignore them with significant consequences. The biggest mistake in the case of ODI is limiting the originating requirements to the voice of the customers. While VOC is adequate for Demand Driven Innovation, it is necessary but not sufficient to Opportunity Driven Innovation as expressed in Fig. 1.3 below:

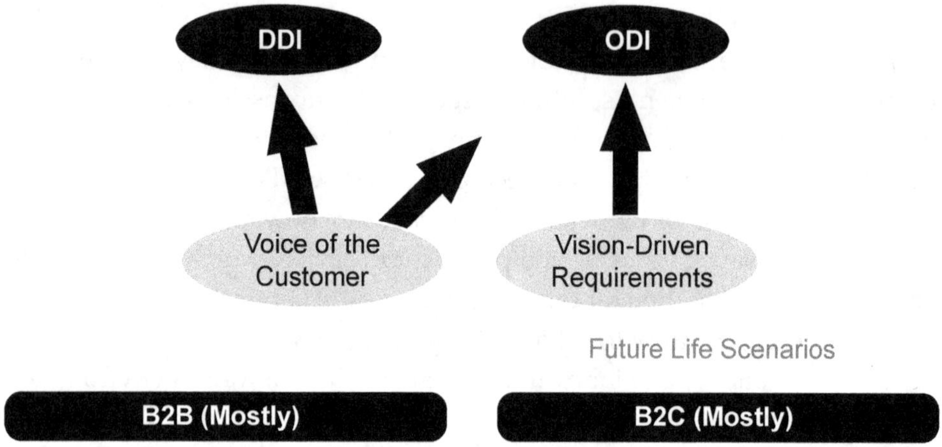

Fig. 1.3 DDI and ODI Drivers

For Opportunity Driven Innovation, understanding the customers' unconscious unmet needs and the functions a system must perform to address them is an essential step toward generating winning solutions. Chapter 6 explores the types of needs and the actions necessary to address them successfully.

1.5 GOALS AND STRUCTURE OF THE BOOK

One of the goals of this book is to share a systematic innovation theory that is universally applicable and simple in its implementation that can help generate innovatively "Must Have" products, services, processes, or even help formulate an effective organization.

In this book, we will take a journey to the heart of innovation. Along the way, we will learn about methods and processes, and tools that will help generate market-leading products and solutions in an affordable manner. It is the journey to Winning Innovation Excellence; in other words, *WinnoveX.*™

Another goal in sharing this experience is to walk systems development practitioners as well as C-suite executives to the core elements of the theory. It is an opportunity to engage both the technical community as well as the non-technical professionals and executives of any company with the same concepts and have them work seamlessly together as one team in serving their customers. By targeting company executives, they can understand the process, support the product development teams, and help make the right decisions to grow their businesses.

The book also covers how companies can design new and innovative products with a shorter time to market by focusing on the elements that drive value. It addresses how to generate innovation driven by customer value or a value-driven innovation of any system.

To better walk along the path to achieving *WinnoveX,*™ the book is structured into two parts for all readers, as shown in Fig. 1.4 below. Additional resources about the execution of specific methods and their results are in the Appendix. Systems analysts, architects and engineers, and other practitioners or broad readers can refer to the Appendix when this sign **ΔP** is displayed.

Fig. 1.4 presents an overview of the main chapters of the book. It shows how *WinnoveX*™ impacts all the stages of systems development as well as the operations' functions involved in the conception, the production, the release, and the delivery to customers.

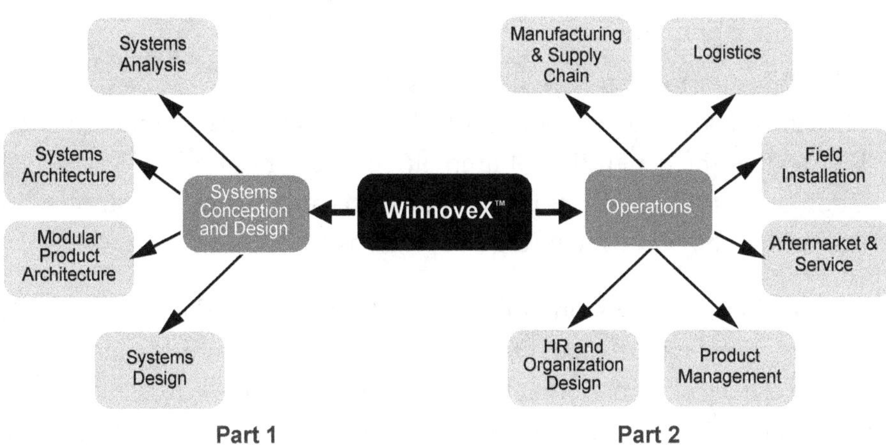

Fig. 1.4 Structure of the Book

Below is a brief description of the chapters of the book.

Part I covers the essential steps of the conception and design of market-leading innovative systems. Chapter 2 introduces the notion of a function through a system analysis leading to the concept of a system's architecture in Chapter 3. The architecture of a system, seen as the integrated structure of its elements and their relationships, has a dual nature. It can be functional (when built from the functions of the system) and physical in its hardware embodiments. Its generic definition has been further refined and simplified in Chapter 4 by defining the building blocks that perform sets of functions and have the property to be tested independently from the whole system. Those blocks are the modules, and the resulting architecture is known as a modular architecture. This property of a module, to be independently tested away from the whole system, drives short development cycle-times, and helps innovate at a faster pace. Chapter 5 explores some details of systems design processes and opportunities with a particular focus on requirement details. No project is successful without detailed and complete requirements.[2] The chapter covers an in-depth analysis of requirements, complete with their allocation from a system to its constituents. Reading and learning chapters 2 through 5 will help understand Chapter 6, which is about achieving *WinnoveX.*™ The application of three essential components proved over several years to work with any system, whether it is a product, a process, a service, or an organization. The three components are functions, technologies, and execution. The resulting method is based on the following principles:

- All entities perform at least one or a set of functions,
- Thoughtful selection of the right technologies to perform the functions is crucial to the generation of innovative solutions and,
- Good execution of both functional analysis and technology application drives the delivery of winning innovations.

So, the simple expression of the method states: "Good execution of any entity's functions with selected technologies leads to winning innovations." That is the crucial statement that the elements of the book will walk through. Chapter 6 articulates the process with essential details. Part II covers Chap-

2 Whether it is a Waterfall or an Agile development process.

ters 7 through 13. It describes how The *WinnoveX*™ Way impacts manufacturing and supply chain in Chapter 7, logistics in Chapter 8, field installation in Chapter 9, aftermarket and service in Chapter 10, product management in Chapter 11, modeling and simulation in Chapter 12, and organization design and human systems in Chapter 13. Chapter 14 highlights a few technologies that deserve some attention since, I think, they will leave their mark on the 21st century. Each chapter ends with an executive summary to allow the readers to quickly grasp the content. It is my ultimate goal to see companies be successful as they travel along *The WinnoveX*™ *Way*. The expectation is that they will gain insight on a path to:

- Launching winning innovative systems consistently,
- Achieving Mass Customization with all its benefits, and
- Emotionally connecting their products, processes, or services to their customers and generate brand recognition and loyalty.

1.6 HOW TO READ THE BOOK

This book intends to reach both practitioners of systems design and non-practitioners professionals and executives involved in the development of systems such as products, processes, service offerings, and human systems such as organizations. Its style is conversational for all readers, and the Appendix covers more technical details for those who may want to dive deeper into the topics. The three suggested reading paths are:

- **Fast read** – Read the Foreword, the Introduction, the Chapter Executive Summaries, and the Conclusion.
- **General read** – Read the whole book ignoring the **ΔP** signaling that there are more resources for practitioners and other in-depth readers.
- **Additional read** – Read the resources available at the **ΔP** sign locations in addition to the whole book (to go deeper).

Part I

The WinnoveX™ Way

Systems Conception and Design

2

Systems Analysis

As shown in Fig. 1.4 of the structure of the book in the introduction, the first step on *The WinnoveX™ Way* is systems analysis. It is the step in which the entity considered is reviewed in detail to understand what its constituents do and how they operate individually and as a whole to deliver the expected output. This chapter describes how the functions of a system are identified, decomposed, grouped, and used to innovate.

2.1 FUNCTIONAL ANALYSIS

The observations of the plant and the butterfly described in the Forewords mentioned their main elements' functions. So, the question to start with is the one below.

2.1.1 What is a Function?

Every human-made device and every living creature perform one or several functions. Simply put, a "function" is what an equipment or, in general, an entity does; it is the task or series of tasks that the entity in consideration needs to perform to achieve the desired outcome or intended objective. Hubka and Eder define a function as "a property of the technical system which describes its ability to fulfill a purpose, namely, to convert an input measure into a required output measure under precisely given conditions." The fascinating detail about such a definition is the mention of the "given conditions." It is essential to keep the conditions in mind when identifying or defining the functions of a system. People familiar with Value Management who use Function Analysis System Technique (FAST) would know the definition of a "function" as "the natural or characteristic action performed by a product or service." Generally, an action expresses a function in the form of a verb and an object. It is also generally advised to use active verbs with measurable functions. For my studies and practice, I use a broader view of the choice of verbs. They need to be verbs with verifiable functions. Such a distinction is important because when we get to functional requirements, we can verify them even though the metric is a binary one such as "On/Off" or "Open/Close." For example, "Open Door" is an active function. Yet there may be no need to be rigorous about how many millimeters the door should be away from its lock to consider it open if the definition is about allowing someone to go through when the door is open. Whatever device operated it to get someone through performed its function, and this is easily verified.

From the two observations mentioned in the Forewords, we have some examples of "Verb and Object" as the expression of a function.

Let us now analyze a simple bottle. When we consider the bottle in Fig. 2.1 below, we can say that its primary purpose is to contain material. Therefore, the bottle is a container designed and produced to hold fluid or loose particles generally. So, the primary function of the bottle is to "contain a fluid." However, if we continue to analyze it, we will identify other essential functions it also performs. We can say that the bottle prevents its content from spilling. The bottle can also have features that enable its grasp to transfer its content. Other features allow us to identify its content. In conclusion, a bottle performs the functions summarized in the table of Fig. 2.1. Those five functions are enough to functionally describe our bottle.

Bottle	Primary Functions	Secondary Functions			
First Level Functions	Contain fluid	Prevent spill	Enable transfer	Enable grasp	Describe content

Fig. 2.1 The functions of a bottle

It is not necessary to decompose them further. We could add the function to "Preserve content," but it is the result of the actions of "Contain fluid" and

"Prevent spill" with the work to seal its content. Functional decomposition, which we will discuss next, is often necessary to better understand all the tasks or desired tasks that an existing device performs, or a future device will perform. It is the best way to understand the functional requirements and their allocation to the elements that make up a system, whether it is a product, a process, a service, or an organization.

2.1.2 Functions Identification and Decomposition

Every system has a "main purpose function," which is also its operational mission function. A car has a mission to transport people and goods. In functional terms, it purely "transports people" or, more generically, "transports load." But while the operational mission function is fundamental in expressing the ultimate value of the system, it is not sufficient to fully describe it. Therefore, other "functions" are needed. Their identification happens through functional analysis.

The functional analysis is the process, mostly iterative, that identifies the functions of a system at various levels and defines their relationships. The process generally starts from the identification of the first level functions (primary and secondary). Then they can be decomposed to lower-level "functions," whose collective actions fulfill them. The exercise derives lower-level functions from the higher-level ones by answering the question: "What needs to happen, or, in other words, which functions need to take place to fulfill the one considered?" The team performing the analysis may or may not identify new functions. The exercise continues until no new "functions" emerge. The ability to perform a sound system functional breakdown is essential to having satisfactory requirements. Identifying the first level functions often happens through a group exercise where system architects and experienced design engineers meet to review what their product does. It often happens that people who have been designing a family of products for many years have difficulty describing what the products do in the concise form of a "function." Therefore, no worries if the team gets stuck first. Several techniques are available to identify the "functions." The practitioners can review the products' operational scenarios or decompose the existing products or simply go through brainstorming techniques. Sometimes several methods are used to get to a comprehensive library of a system's functions.

Considering the car example again, its higher-level function "Transport people" can be decomposed into:

- Load people
- Move people
- Protect people
- Unload people, etc.
- Etc.

The decomposition of each of the functions could continue further. Once higher-level functions are decomposed into their lower level functional components, the resulting structure will look like Fig. 2.2.

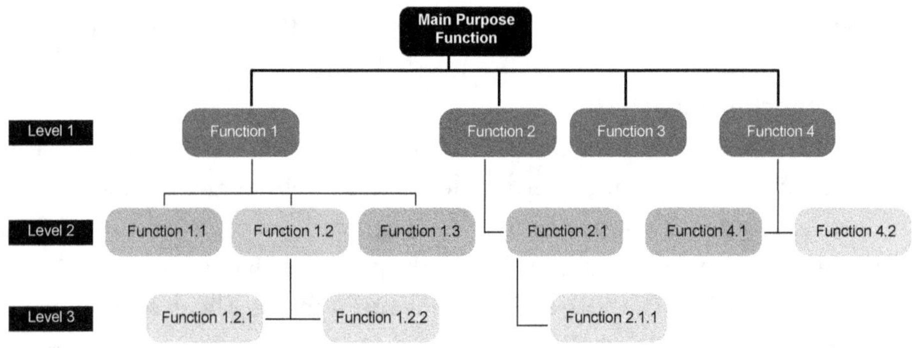

Fig. 2.2 Functional decomposition result structure

Before diving into functions and all their implications in the next chapters, it is essential to explore some techniques that will lead to their determination.

a. Operational Scenarios Review (OSR) Method

Operational scenarios are a succession or sequence of actions that any person interacting with a system would take. Therefore, there are several scenarios relative to users, assembly process, regular maintenance, repair, emergency intervention, etc. The sequence of the actions must be carefully captured and described in a document. The operational scenarios document is an essential reference for how the system operates or should operate.

Once created, its use identifies the actions taken by the operator with the resulting system's behavior. The next step is to translate the actions and behaviors into the functions the system performs in response to an operator's input.

Let us consider an elevator operation. The example illustrates a user scenario with a passenger taking an elevator to go up or down. The process is called in elevator jargon, "a single trip normal run." The passenger interacts with the elevator by giving input to the system, whose response shows some visible signs while taking other actions behind the scenes. Table 2.1 shows the summary of a single trip normal run. It starts with the passenger calling the elevator by pressing a button.

Step	User Visible Behavior	System Behavior	Functions
1	Passenger presses Hall button		Request service
2		System recognizes service request	Acquire request
3	Hall button "tell-tale" light activates		Acknowledge request
		System assigns a car to service request	Dispatch elevator
		System opens the brakes	Move car
		Car travels to service request location	Move car
4	Hall light activates	System activates hall light	Indicate arrival
5	Car door open	Car stops at service request location	Stop car
		Drop the brakes	Stop car
		System opens doors	Allow access
6	Passenger enters car	Keep doors open	Allow access
7	Passenger presses button in car		Select destination
8	Car button light activates	System recognizes destination selection	Acknowledge request
9	Car and hall doors close	Secure passenger in the car	
10		System opens the bra	
	Car moves	Car tr	

Table 2.1 Elevator user scenario for a single trip normal run

ΔP The full table is in Section A1.1, Appendix A.

In the end, the exercise identifies several functions, as shown in Table 2.2.

(*) Each subsystem contains several groups of functions that fulfill its primary function. The notion of subsystem will be explained more in Chapter 3.

As first-level functions in Table 2.1, they would need further decomposition. At the right level, their association with a metric would define a

Functions	Subsystem (*)	Comments
• Request Service	• Human Interface	
• Acquire Service	• Controls	
• Acknowledge requesr	• Human Interface	
• Dispatch elevator	• Controls	
• Move car	• Propulsion	
• Indicate arrival	• Human Interface	
• Stop car	• Propulsion	
• Allow access	• Doors	
• Select destination	• Controls	
• Protect load	• Vehicle	
• Inform passenger	• Human Interface	
• Allow egress	• Doors	

Table 2.2 First level functions generated from the single trip normal run scenario

functional requirement by expressing how well their performance achieves the expected outcome.

Functions such as "dispatch elevator" or "protect load" are performed only when several lower-level functions happen either sequentially or simultaneously.

For an elevator in a building, the "single trip normal run" is just one of many scenarios. The same elevator will perform other functions when a firefighter (in case of an emergency evacuation) or a passenger has a keyed operation or when mechanics rides part of it during its installation. The operational scenarios to identify the functions continues with one after another, followed by further decompositions until the team reaches the lowest level (elementary) functions. For the case of an elevator, the following are the scenarios to consider:

- Passenger scenarios (single run, keyed operation, sabbath operation)
- Installation scenarios
- Maintenance scenarios
- Repair scenarios
- Emergency operation scenarios
- Commissioning scenarios
- Customer expected scenarios (need creativity to be determined and may involve the use of focus groups)

The OSR method opens up the opportunity to consider new scenarios beyond the known ones. Therefore, after reviewing all the already documented cases, the process continues beyond existing product designs to uncover new use cases. It provides a direct way to identify a system's first level functions, whether existing or new. The hard work is to document the scenarios. It is a team exercise that requires domain experts' involvement. They have the most knowledge of the context in which the system operates. However, they may not be used to thinking in terms of functions, so an experienced facilitator could be very helpful. The use of tables like Table 2.1 shown above provides the framework in which the description of an operator's behavior and the system's response lead to a function or set of functions. While most of the scenarios may lead to identifying the functions performed by systems that are already commercially available, new use cases can also emerge. In that case, it requires the ability to anticipate the actions not currently used by the users but could become part of their habits. Companies or design groups that invest in researching new user scenarios and introduce new functions are considered innovators. Chapter 6 covers the topic. The key message here is about introducing innovation, either through new ways of performing existing functions or through the offering of new functions.

Here is an elevator innovation that is becoming popular. Users usually push a button in front of the elevator entrance to call the car. In functional terms, they request a service (step 1 in Table 2.1). Once inside the elevator, the passenger selects a destination by pushing a button on the car operating panel (step 7 in Table 2.1). This is usually how elevators work today. The elegant innovation consisted of allowing the passengers to "select the destination" before entering the cabin at the time when they "request the service." The system acknowledges the service request by directing the passenger to the right elevator. Once the system performs the two functions simultaneously, the cab is sent to the correct floor while the passenger, as service requestor, is walking toward the elevator. The new solution saves the uncomfortable waiting time people experience in front of an elevator's closed door. Several devices, such as calling the elevator using employee badges from turnstiles or using key fobs or remote stations, or even mobile phones, are now in use. Some elevator companies have gone so far as to eliminate the operating panel from the cab; so, with the right approach, a system's functions can also reduce its cost.

In addition to the Operational Scenarios Review (OSR) Method, another way to identify a system's functions is through the breakdown of existing products.

b. Existing Products Breakdown Analysis (EPBA) Method

The EPBA method may be the one understandable to most people. It leverages a company's existing products whose constituents are known as part of various configurations. The process essentially consists of three steps:

- **Step 1:** Decompose the existing products into their building blocks
- **Step 2**: Determine the functions performed by each of the blocks
- **Step 3**: Regroup the functions from the lowest level parts, without the constraint of how they were performed by the existing products, to a level where they can now be performed by various technologies, using the new groupings to create the functional architecture.

Functions from the lowest level parts exist because of the previous design and implementation decisions. The existing functionality needs to be brought to a level 'above' where technology and allocation decisions were locked in to innovate with different technologies and implementations.

ΔP Detail of the steps is in Section A1.2, Appendix A.

The example of the simple bottle considered earlier offers the mapping represented in Fig. 2.3. Generic elements, both functional and physical, perform the bottle's functions. Being functional, they are independent of any technology implementation. And they are also physically recognizable elements. The next sections and chapters will cover the dual functional and physical nature of the newly defined elements.

Bottle	Primary Functions	Secondary Functions			
First Level Functions	Contain fluid	Prevent spill	Enable transfer	Enable grasp	Describe content
Generic Element	Container	Cap	Container	Handle	Label

Fig. 2.3 Bottle's elements and their functions

The goal of the exercise is to go from known physical entities to functional ones and their relationships.

The question that often comes up is, why not proceed directly into the decomposition of the physical products and decide their subsystems and modules on such a basis? By doing the straight breakdown of existing products with no regard to the functions their elements perform, the resulting systems are strictly part of the specific technologies that generated them. That is why I emphasized earlier in step 3 to regroup the functions at a level above where the previous design teams made their implementation decisions. If not, it becomes harder to think independently and beyond the product's existing design solutions. Also, the physical decomposition of a system based solely on existing technologies does not pass the test of time due to technology obsolescence. The same example of a bottle shows that the need to "contain a fluid" still existed even when bottles were no longer made in animal skin but, instead, with various other materials. By going into the functional domain, the design teams can consider many different "How's" to perform the functions, which gives them more flexibility to innovate based on a whole variety of technologies. While technologies evolve at a relatively faster pace, functions change very slowly, opening the opportunity to a continuous flow of various technologies to perform them. Therefore, the functional architecture of a product becomes the anchor for innovation.

A more complex example will further illustrate the method and its outcomes.

Example: City and Motorized Bicycles

A bicycle is a popular system. Therefore, it makes an excellent example to illustrate the EPBA method.

Fig. 2.4 shows a few bike types, including a regular city bike, a motorized bicycle, and an electric bike.

Fig. 2.4 Bicycle Examples

(Sources: amazon.com; ancheer.shop, and jaguarpowersport.com)

A hierarchical representation of the breakdown of the bikes' main legacy elements and sub-elements is in Fig. 2.5 due to their observation. It is not fully representative of a specific manufacturer's product. It includes details

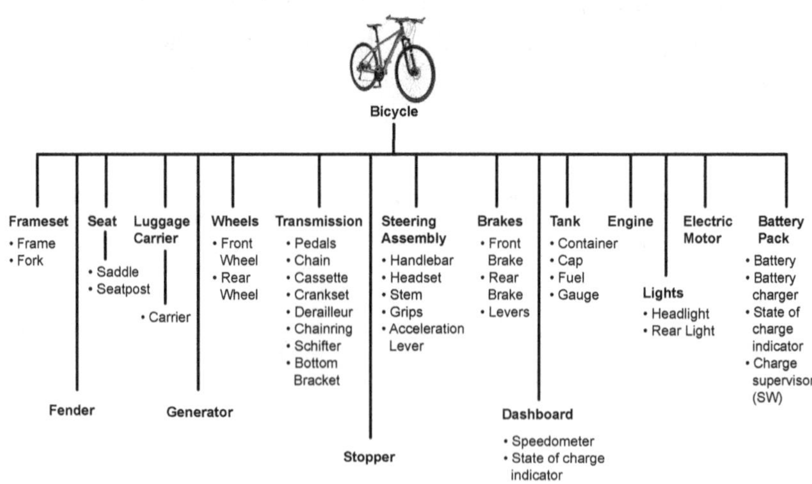

Fig. 2.5 Bicycle physical breakdown

from many bikes, such as a regular city, motorized with a fuel engine, or electric motor.

Once the initial physical breakdown is complete, several steps follow. The next step is determining the first level primary and secondary functions of each element and their further decomposition if necessary. Table 2.3 shows the result of the effort where high-level constituents are broken further down into their components. The physical domain is about the bike's hardware with any software loaded onto them. The functional domain matches the physical by identifying the functions performed by the elements at their respective levels of the hierarchy.

Physical Domain		Functional Domain	
Product Level 1	**Original Physical Element (Design Solution)**	**Functions**	**Lower Level Functions**
Frameset		**Carry Load, Transfer load,**	
	Frame		Distribute load, Hold wheel,
	Fork		Transfer load, Hold front wheel, Absorb shock
Seat		**Enable seating**	
	Saddle		Provide seat, Carry load
	Seatpost		Transfer load, Hold seat, Enable seat adjustment
Luggage carrier		**Carry load**	
	Carrier		Carry load
Front wheel		**Enable motion**	
	Front wheel		Enable motion, Provide braking surface
Transmission		**Transmit motion**	
	Pedals		Generate torque, Start motion, Sustain motion, Rest feet, Propel bike, Reduce speed
	Chain		Provide power, Transmit motion, Vary speed
	Cassette		Select speed, Change speed
	Crankset		Convert motion, Tran[illegible] Cha[illegible]
	Front Derailleur		
	Rear Derailleur		
	Shifter		
	Bottom bracket		

Table 2.3 Bicycle breakdown with the primary and secondary functions

ΔP Full table is in Section A1.2, Appendix A.

All the identified functions will be performed by "Modules" that represent the portfolio of the bikes shown in Table 2.3 above. The modules are then grouped under higher-level elements to form "Subsystems," as shown in Table 2.4. Chapter 3 covers both module and subsystem concepts in more detail.

What is important to note here is the transition from the bikes' hardware (physical entities), decomposed into their constituents to extract their functions (functional entities), which are then regrouped and assigned to new generic entities. Such a step is critical to the success of future products' design because it gives the designers all the freedom they need to apply any appropriate technology available to perform the functions and deliver the value of the product. I cannot emphasize enough the passage from physical (existing parts and components) to functional entities before deriving the modules with their dual nature both functional and generic physical. They are physical entities having well defined functional expectations in addition to specified interfaces. They are both functional (because they represent a set of functions and without any implementation technology) and physical (because they embody hardware during the design phase). Table 2.4 is the summary of the effort.

Physical Domain		Functional Domain		Modules Functional and Physical	Subsystems
Product Level 1	Original Physical Element (Design Solution)	Functions	Lower Level Functions	Functions to Physical Mapping into Modules	Modules Grouping to Subsystems
Frameset		Carry Laod, Transfer load,			Frameset
	Frame		Distribute load, Hold wheel,	Frame	
	Fork		Transfer load, Hold front wheel, Absorb shock	Fork	
Seat		Enable seating			Load Carrier
	Saddle		Provide seat, Carry load	Seat	
	Seatpost		Transfer load, Hold seat, Enable seat adjustment	Seatpost	
Luggage carrier		Carry load			Load Carrier
	Carrier		Carry load	Luggage Carrier	
Front wheel		Enable motion			Wheels
	Front wheel		Enable motion, Provide braking surface	Front wheel	
Transmission		Transmit motion			Propulsion
	Pedals		Generate torque, Start motion, Sustain motion, Rest feet, Propel bike, Reduce speed	Pedals	
	Chain		Provide power, Transmit motion, Vary speed		
	Cassette		Select spe		

Table 2.4 Bike Functions, Modules, and Subsystems

ΔP Full table is in Section A1.2, Appendix A.

We can now consider the functional building blocks of the portfolio with the subsystems and modules, as shown below in Table 2.5.

SS ID	Subsystems	Mod. ID	Modules
1	Frameset	1.1	Frame
		1.2	Fork
		1.3	Fender
2	Propulsion	2.1	Pedals
		2.2	Motion transmitter
		2.3	Speed selector
		2.4	Motion converter
		2.5	Power converter
3	Wheels	3.1	Front wheel
		3.2	Rear wheel
4	Bike Control	4.1	Steering assembly
		4.2	Acceleration laver
		4.3	Stopper
5	Brake	5.1	Front Brake
		5.2	Rear Brake
		5.3	Brake cable
		5.4	Brake levers
6	Dashboard	6.1	Speedometer
		6.2	State of charge indicator
7	Load Carrier	7.1	Seat
		7.2	Seatpost
		7.3	Luggage carrier
8	Power Generation	8.1	Generator
		8.2	Engine
		8.3	Tank
		8.4	Fuel
		8.5	Electric motor
9	Energy Storage	9.1	Battery pack
		9.2	Battery charger
		9.3	Charge supervisor (SW)
10	Lights	10.1	Headlight
		10.2	Rear Light

Table 2.5 Bicycle Functional Structure: Subsystems & Modules

Fig. 2.5 is redrawn based on Table 2.5, with the newly defined subsystems and modules. The new representation is in Fig. 2.6.

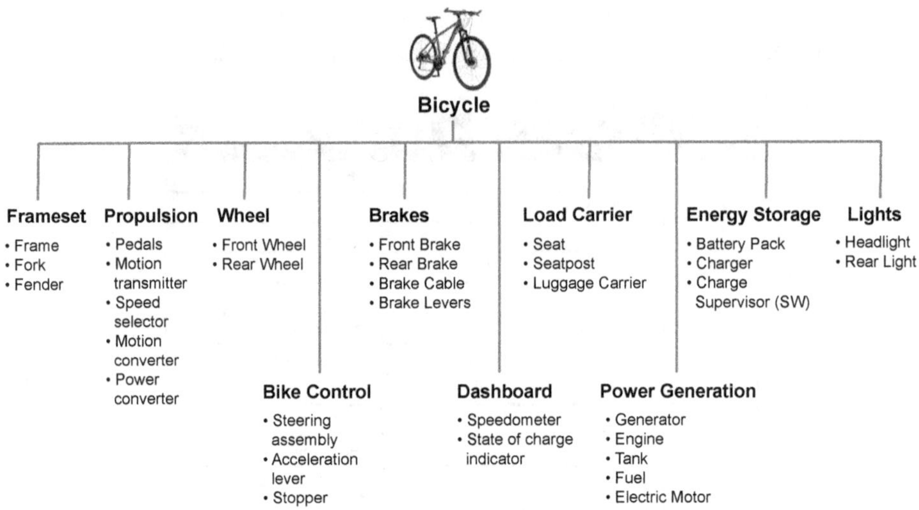

Fig. 2.6 Bicycle Functional Structure

ΔP Another representation is in Fig. 2.7, shown in Section A1.2, Appendix A.

There are other ways to get to the functions performed by a system.

One is a variant of the EPBA (Existing Products Breakdown Analysis) method. It is the Operand Flow Analysis (OFA) method. The OFA method is a subset of the EPBA methods and a way of decomposing existing products into elements that lead to the definition of subsystems and modules. An operand is any element that flows through a system and interacts with several of its constituents. Examples of operands are blood for the human body, air in an air compressor, or electricity in many machines. The method assumes that by following the operands, the functions performed by the system's elements they interact with, emerge.

ΔP A fully detailed example of the OFA method applied to a Vapor Compression based HVACR[1] system is in Section A.1.3 of Appendix A.

1 HVACR = Heating, Ventilation, Air Conditioning, and Refrigeration

The illustration shows the definition of a Vapor Compression Based Architecture (VCBA) of an HVACR system.

Another one is known as the Function Analysis System Technique or FAST. It is a method that is used in Value Engineering to define the functions performed by a system. Its use of functions is interesting to know. Details about its application are in literature. Therefore, I will not dwell too much on it. But I will just mention that the method uses an intuitive logic to check the functions. A graphical model is usually used with the functions horizontally from left to right are in a How and Why relationship.

ΔP A few words about FAST are in section A.1.4, Appendix A.

2.1.3 System Functions Grouping

The ability to identify an object's functions is one of the most critical product design's steps. Coupling functions with how well they are performed, they define the Functional Requirements, which are the starting point of any design. It is of prime importance to identify the functions of an entity and to group them, as explained earlier, into sets that facilitate synthesis, i.e., product design. Just like in the physical world, any equipment is decomposed into subsystems, which are made of modules, then of components, and ultimately of parts, functions can also be grouped into functional parts, functional components, functional modules, and functional subsystems. Such groupings are functional components, functional modules and functional subsystems. The notion will be explained in depth in Chapter 3. Functions' grouping must follow principles that facilitate the assembly of their physical counterparts with solutions based on independent requirements. When functional requirements have dependencies, not only the resulting design is costly, it is also sub-optimized. If two modules A and B have requirements that depend on one another, they must be tested together to get meaningful results. By extrapolating, if all the modules of a system are dependent on one another, none of them can be designed and tested alone, and the whole system must be built and tested as one entity. It is a costly practice and raises more issues during any eventual troubleshooting. Therefore, Nam Suh's

Independence Axiom, according to which one should maintain the independence of functional requirements, becomes a critical principle to adopt.[2]

In the previous sections of this chapter, functions and their identification process were at the center of all the attention needed to grasp the notions introduced. The value of a system is usually related to its functions and its cost. Therefore, Value Engineering experts have defined value as the ratio of Function over Cost. This definition of value leaves out a critical concept raised by Donald A. Norman called *Emotional Design*.[3] The author expressed how emotions and cognition influence our decisions in different ways. While the cognitive part of our being leads conscious and rational decisions, the emotional side happens in the unconscious. As the author puts it, "Consciousness comes late, both in evolution and also the way the brain processes information; many judgments have already been determined before they reach consciousness."[4]

A product, a public infrastructure, an artwork, or even an organization, have something that drives their perceived value beyond the functions they perform. This nonfunctional element determines their appeal. It triggers an emotional connection to them. In this book, I associate it with the concept of Emotional Resonance, for which I have developed the Emotional Resonance Index (ERI).

2.2 EMOTIONAL RESONANCE AND INDEX

2.2.1 What is Emotional Resonance?

The ultimate goal of any product, service, process, artwork, etc. (systems in general) is to improve people's lives by either:

- Reducing their effort in doing something, or
- Enhancing or augmenting their being and their capabilities.

Generally, well-conceived systems achieve one or both objectives. In the process, the effort leads to understanding the potential customers' unmet

2 Nam P. Suh, *The Principles of Design*, Oxford University Press 1990, p.9

3 Donald A. Norman, *Emotional Design*, Basic Books 2004

4 Ibid, p.11

needs,[5] and companies deploy their resources to satisfy them. The design process involves requirements with rigorous metrics designers must meet through well-established means. But as I mentioned earlier, there is something that goes beyond the technical aspects of a product or, in general, a system. For example, people may be used to taking an airplane to travel. They may not pay any attention to the planes anymore due to all the airport protocols and other distractions. It may even become a routine, but when the Concorde was either taking off or landing at any airport, everybody rushed to see it. During my trips to Paris, other pilots would even announce it, and they always get every passenger's attention. What made the Concorde a symbol went beyond its look or the technical marvel it represented or its nature as the only supersonic passenger plane.

Like the Concorde, other human realizations produce the same effect on people. The influence can touch many people globally or a small group or even discrete individuals drawn toward a product to become loyal to its adoption. It is the topic described in this section. In recent years, maybe due to the rapid development of digital products, two words have become part of design teams' vocabulary worldwide. They are User Experience (UX) and User Interface (UI) or, more generically, Human Machine Interface (HMI). Emil Lamprecht[6] wrote an excellent article for Careerfoundry to explain the difference between the two disciplines and the professions that go with them. UX and UI are making significant advances. UX focuses on the user's overall experience with a product from its purchase to its use. It is about how a user feels that will determine her or his decision to continue buying the product in the future or to recommend it to someone else. Several publications feature UX methods. UI/HMI, on the other hand, enhances the look and feel of the product through its interface with the user. As Steve Krug puts it, "Don't make me think."[7] He presented the principles of intuitive navigation and information design. The intentional management of the two fields is still expanding even though the great designers of the past and present seem to grasp the UX aspect of their creations intuitively.

Through my observations and conversations with people from various regions and cultures worldwide, I reached one conclusion. While both UX

---

5 The topic of the determination of the unmet needs will be covered in Chapter 6

6 Emil Lamprecht, CEO of Growth Mechanics

7 Steve Krug, *Don't Make Me Think*, New Riders, 2014

and UI are essential and necessary drivers in establishing users' connection to a product, there seem to be other factors that, along with them, reach people at their core, originated from their unconscious inner self. Products that successfully carry these factors along with well-established UX and UI put people in a state of Emotional Resonance. It means that the product emotionally resonates with the users or, in general, the customers. It creates an emotional bond with whoever gets in that state. They become advocates of the creation and admirer of the author or producer, whether it is a person or a company.

During my student years in Italy, I spent some of my Summers as a tour guide in Rome. I was fascinated by the attention people paid to the history of the monuments. Later as a professional with the opportunity to work with teams worldwide, I took every chance to ask people what motivated them to make choices like a specific trip, a purchase, a visit, etc. It is from such interactions spanning three decades that I concluded the existence of factors that determine the Emotional Resonance. So, the question became how to understand them and formulate a way to trigger them intentionally.

2.2.2 Elements of the Emotional Resonance Index (ERI)

Compared to a function, which Hubka and Eder[8] defined as "a property of the technical system which describes its ability to fulfill a purpose", Emotional Resonance, in the context of this work, is a property of the system to connect with the inner unconscious emotional side of people as they interact with it. It leaves a mark in them and triggers their desire to own or be part of the system's "life." Emotional Resonance has both objective and subjective variables. To express it, I used the Emotional Resonance Index or ERI. ERI is made of several elements and defines the character of a system and its lasting emotional connection to those who own or use it. Two similar systems may perform the same set of functions, and yet one could be perceived by the customers or users as more valuable and deserves a higher price. Understanding the Emotional Resonance Index of a system could be crucial for its commercialization and be explored as the predictor of an impactful system.

When a wide range of several thousand people were randomly asked, over more than two decades, in front of stores and in big tourist cities world-

8 Vladimir Hubka and W. Ernst Eder, *Theory of Technical Systems*, Springer-Verlag Berlin, Heidelberg 1984, p.72

wide why they bought a given product, or decided to visit a specific place, or stopped to take a picture of a piece of infrastructure, they were unanimous on some answers. Their reactions led to the definition of some factors and observation of some behavioral steps. The most familiar, widespread, and consistent comments were around aesthetics, perceived performance, and charisma.

- **Aesthetics**. For many, the word is synonymous with beautiful. We would agree that "beauty" is subjective and relative to individuals or a social group of people and their culture. However, some features seem to transcend culture, geography, gender, generation, etc. Everyone appreciates them and unanimously admit that the system that embodies them is beautiful. Those were the features of this study's focus.
- **Perceived performance**. It is not a system's established performance based on defined, measurable metrics. It is a "Perceived Performance." Anybody that buys a device, or interacts with a product, immediately develops some expectations. A product designed with rigid metrics for the designers to meet still has its final judgment from those who interact with it. It depends on how they perceive it through their senses or their mental image of what it should be. People need to feel like the system was a part of them as if it was enhancing their persona or augmenting them. Their experience is enjoyable, and if it is a product, it fulfills the job to be done working well and presents no threats. It is a feeling, a perception. For example, everyone knows that a plane in turbulence will go through it without falling from the sky; and yet the discomfort makes it a bad experience. Among the attributes that impact the perception of a good or bad performance, we could mention noise, vibration, heat, cold, smell, brightness or darkness, edges (round or sharp), flimsiness, order (or disorder). Any perception of abnormality of one or more of them, while interacting with the system, puts people in an alert mode. No matter how great a design is, the perceived performance could still be a bad one.

Interestingly, even though perceived value came out of a general observation, as human perception, it is still subjective and relative to each person. Therefore, its measures cannot be absolute. Its characterization must happen through surveys and focus groups whose results should include local cultures. I discovered, through my interactions, that people express Perceived Performance with terms like, "it works all the time," "it is safe," "it is reliable," "it is worth the price," "I enjoyed it," "The experience was worth doing," "You should get it," etc.

- **Charisma**. As defined by the Cambridge Dictionary, charisma is the superior power natural to some people that makes them able to influence other people and stimulate their attention and admiration. Another dictionary defines as charismatic, a person exercising a compelling charm that inspires devotion in others. Charisma's Thesaurus synonyms are "attraction," "glamorousness," "magnetism," etc. By extrapolating from people, in terms of charismatic leaders, I see that engineered systems can also exhibit a charismatic appeal on people. It makes people want to be associated with them. Take, for example, a Ferrari car; owners always come out with a glow coming from their possession. They can feel everybody's admiration for their vehicle. Architects know how to design their buildings with charisma. In the survey that sets the charismatic level of any creation, people, when asked how they feel about the system of interest, they just go on and on about it. It is electric. When the feeling, the connection, the desire to acquire it, the readiness for some sacrifice to own it, or defend it, are absent or lukewarm, it means that the level of charisma emanated by the system is low.

Teams can assess the ERI of an existing system as well as one under development or future. A survey of customers and users is vital to determine their level of engagement leading to loyalty. The Index is not an absolute, but a directional measure to help product developers and service providers avoid unsuccessful releases. Surveyors of existing products or services must reach the people at their point of purchase or use. They can get direct feedback at a store when they make their purchase or meet at a place where they interact with the product. Such transactional surveys combined with long term users' feedback are an essential input to the process.

For products or services that are still in development, proof of concepts, prototypes, or digital means such as virtual interactions are ways to engage the customers. Their feedback will provide meaningful results if they understand the product or service and the needs that they will satisfy.[9]

9 The various types of needs are described in Chapter 6.

2.2.3 Emotional Resonance Index

Assuming aesthetics as beauty characterization, perceived performance as a feeling of excellent performance, safety, comfort, etc., and charisma as magnetism, attraction, an inspiration for being more, etc., the definition of the Index is as follows:

- Each of the variables is on a scale from 1 to 10 expressed through surveys and focus groups,
- Three cubes defined as the "Emotional Resonance Index Cubes" are as shown in Fig. 2.12,
- The inner cube has a side length of 7, the middle cube a side length of 9, and the outer cube has a side length of 10 with no particular unit. The volumes are 343, 729, and 1000 respectively.
- Each side length number is defined as the average rating of all the people interviewed for each variable, aesthetics, perceived performance, and charisma.
- The Emotional Resonance Index (ERI)[10] is defined as the volume of a cube whose side length is equal to the average rating of each variable described above.

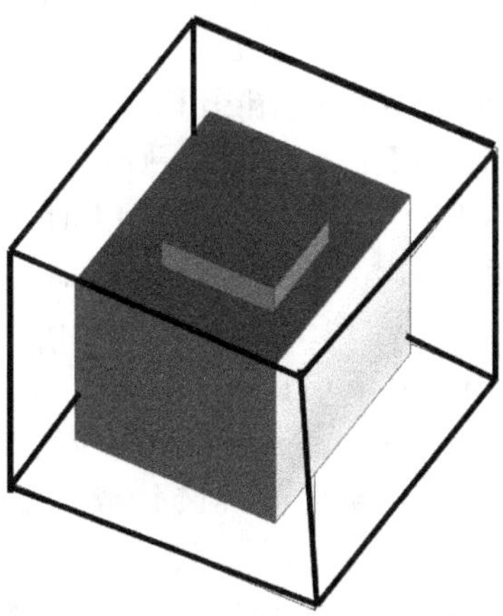

Fig. 2.12 Emotional Resonance Index Cubes

» When the ERI is between 729 (medium cube volume) and 1000 (outer cube volume) and none of the three variables scores less than 9 (side

10 The other figure numbers of Chapter 2 are in Appendix A.

length of the medium cube), the product, service, process, or in general, the system is likely to generate loyal customers or users that will continue to buy or use it and recommend it to others.

» When the ERI is between 343 (inner cube volume) and 729 (medium cube volume) and none of the three variables scores less than 7 (side length of the medium cube), more work must be done to improve the scores and the Index.

When the ERI is less than 343, the system is not likely to generate any emotional resonance.

A few words about a couple of surveyed products would provide a better sense of the ERI.

Even with wristwatches introduced more than a century ago, Apple Inc. managed to surpass famous historical names to launch its Apple watch. The technologically advanced product was not the first electronic wristwatch introduced on the market, but it still scored high for aesthetics, perceived performance, and charisma. After its introduction, as I do to other products, I had casual conversations in various locations worldwide with people who were wearing it. They always get shiny eyes when asked if they like their watch. After the launch of several generations of the product, it is still noticeable to see users boasting about it. The following generations of the Apple watch continue to score a very high Emotional Resonance Index.

Another remarkable high ERI scorer is Leonardo da Vinci's Last Supper painting, in the Convent of Santa Maria Delle Grazie in Milan, Italy. The 15th-century painting still has its daily long queue of patient visitors who wait for their turn to see it. Every time I have the opportunity to travel to the city, I visit the place. It is impressive to see people come out sharing how the Italian Renaissance mural painting emotionally resonates with them. The artwork is one of the biggest attractions in Milan.

Emotional Resonance creates a bond between a product, or in general a system, and the customers. Any design should assess it and pursue it on top of the system's functions and the technology choices to create long-lasting customer loyalty.

2.3 SYSTEMS ANALYSIS SUMMARY

Every entity, whether natural or engineered, performs a function or a set of functions. A function is what the entity does or the task(s) it performs to fulfill its purpose. A bottle's primary function is to contain fluids or loose particles. Understanding the functions of an existing or a new system, whether it is a product, a process, a service offering, or an organization, is critical to its design. Once a design team identifies or defines a system's functions, it can select the appropriate technologies to perform them. Technologies evolve and become obsolete, but the functions they perform remain mostly current. They express the needs whose satisfaction brings value to customers. For example, nobody carries water in animal skins anymore, but the need to "carry" it is still present. There are several thousands of bottle designs with various technologies that are on the market with the function to contain fluids and satisfy the need to carry water. The knowledge of an entity's function brings total freedom to rethink it and to redesign it by leveraging new ways or technologies to enhance its value. Technologies enable the translation of functions into real and useful systems. An entity's functions identification or definition process is known as "functional analysis." The chapter covered several functional analysis methods with examples of how they work.

The chapter also covered the notion of Emotional Resonance with its assessment Index (ERI) for a system (product or service). ERI is a property of a system to connect with the inner emotional side of people as they interact with it. Its Index is the measure of a customer's appreciation for the aesthetics, the perceived performance, and the charisma of the entity in consideration. Understanding the Index for an existing product can help innovators make adjustments to its design or its marketing. For a new product or service, it can enable designers to make the right decisions from its conception. Overall, understanding the Index can make a difference in adopting a product, a process, a service, or even an organization as a human system. ERI assessment is vital for new designs and existing products to create bonds that drive customer loyalty.

Through systems analysis, companies have the opportunity to clearly understand the functions their products and services offer to satisfy their customers' needs. Such knowledge can help them make the right technology choices that differentiate them in the marketplace and keep them on the leading edge.

3

Systems Architecture

Chapter 2 introduced the notion of a system's functions with various methods to identify them and regroup them. In this chapter, we will revisit the grouping of functions and their implications for designing any type of system. We will go into more detail about the concepts of module and subsystem and the whole hierarchical structure in which they represent a system.

3.1 DEFINITIONS

Before introducing the notion of systems architecture with all its implications, it is crucial to define a few terms in the context of this book. In the hierarchy of the elements of a system, there is a correspondence between physical and functional entities.

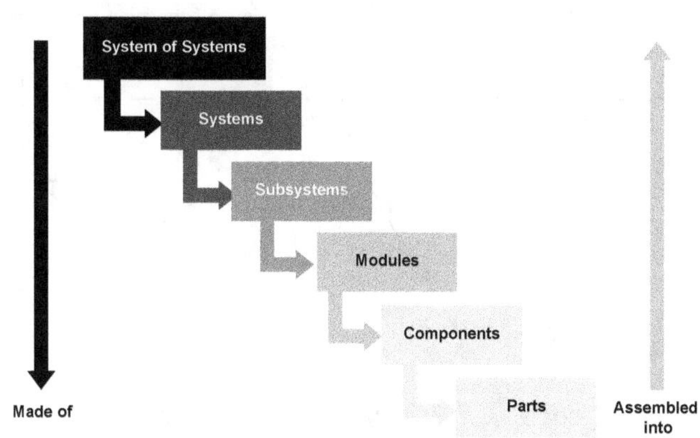

Fig. 3.1A Hierarchy of system elements

Understanding the parallel and relationship between the functional and the physical domains leads to better use each of them. Fig. 3.1 shows the hierarchy of a system's elements as used in the book.

As shown above, top-down,

- Systems of systems (SoS) are made of Systems
- Systems are made of Subsystems
- Subsystems are made of Modules
- Modules are made of Components and
- Components are made of Parts

Bottom-up, lower-level elements' assemblies make higher-level elements. In some cases, especially with large systems, we could have sub-components between components and parts, but we will proceed with the hierarchy that is common to most systems.

The same hierarchy can be viewed in a more usual way, highlighting the elements and their links, as shown in Fig. 3.1B Systems A, B, and C form together, a system of systems. Each system is an assembly of subsystems broken into modules and subsequently into components, broken into parts.

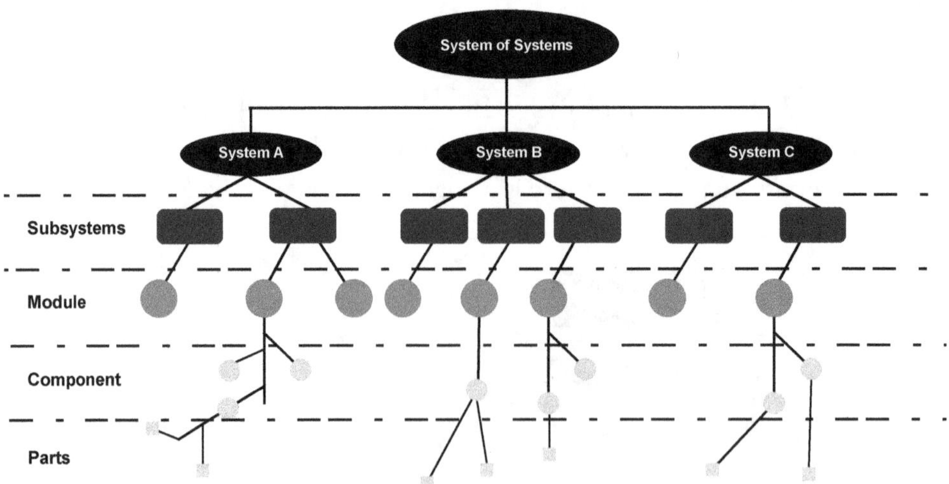

Fig. 3.1B Hierarchy of System Elements

In the following sections, the system elements of Fig. 3.1A and B are further defined as used in the book.

In systems literature, there are many definitions of each element. Therefore, it is essential to state what they mean in the context of this book. Chapter 2 already covered some of the notions, but without the depth of their definition. They will be appropriately defined in this chapter.

3.1.1 System and System of Systems

Generally, the term System applies to any entity made of integrated elements forming a unified whole with a purpose not achievable by any of its constituents. The term "integrated" is vital as it expresses the relationship or interaction of the elements to work together to achieve the purpose of the whole entity. It is different from an aggregation that is a simple collection of disparate elements without the expectation of fulfilling a joint or common outcome. A house, an automobile, an airplane are a few examples of systems. Systems are often combined to form a system of systems. In a city, the transportation system of systems is the assembly of vehicles system (cars, buses, subways, tramways, etc.), roads system, traffic light system, etc. In this book, only the term system will be often used even for systems of systems. Beyond natural systems, human-made systems are designed to perform an "operational mission" function.

3.1.2 Physical and Functional "Subsystem"

A subsystem is a subordinate group of elements of a system. A system could itself be a subsystem in a larger system. In the functional domain, the equivalent of a subsystem is also called a segment. But in the book, the choice is more on using "Functional Subsystem." When the word "Subsystem" is used without the functional qualifier, it means a physical subsystem. As subordinates of the whole, subsystems interact to allow the system to meet its requirements. The powertrain and the body of a motor vehicle are examples of subsystems.

3.1.3 Physical and Functional "Module"

In this book, a module is an independent assembly designed to perform a set of functions and deliver an expected output. Such an output depends on a given set of inputs under specific conditions. In chapter 2, we reviewed the notion of function grouping. The exercise of grouping functions must avoid any dependency between them [Section 2.1.3]. Independent function groupings also assure the independence of the modules that embody and perform them. As introduced in Section 2.1.3, function sets are also called functional modules. A functional module is equivalent to a generic physical entity that represents all its functions in the physical domain. We will use indifferently "functional module" and "generic physical module" with the same meaning. As explained in the next Section 3.2, the container of a bottle as a module can perform other functions beyond "contain fluid", such as "enable transfer" and "enable grasp." The function set performed by the container module will be made of all three functions:

- Contain fluid
- Enable transfer
- Enable grasp

The container module is, therefore, functional (because it performs a set of functions). Still, it also represents a generic physical entity that contains fluids and can be grasped to transfer its content.

Well-defined and designed modules can be tested away from the rest of the system to confirm their output based on their input. Modules can be mechanical, electrical, or software, or a combination. In a bottle, other modules beyond the container, would be the cap, and any label attached to the container [Fig. 2.3]. An analogy to the human body as a system would identify an organ as a module.

With their property of independence and capability to be tested alone, modules have a life of their own and can be re-used across multiple systems. Such a property makes them the building blocks of a system.

Special attention is given to the way modules connect or interface to one another. The definition of a module's interface requirements is one of the most critical steps of systems design. Sections 3.5 and 5.2 cover the topic. A module's interface drives its ability to:

- Perform its group of functions and meet its design requirements,
- Be tested without the whole system,
- To be re-used in multiple systems.

The importance of an interface management was on full display in the medical field when Dr. Chris Barnard opened new possibilities on December 3rd, 1967. He performed the world's first human heart transplant at Groote

Schuur Hospital in Cape Town, South Africa. In this case, the mechanical interface of the heart is made of the connections to the major vessels that bring oxygen-rich blood to the heart and those that take oxygen-poor blood from it. We could name the pulmonary veins, the coronary arteries, the superior vena cava, the inferior vena cava, the pulmonary artery and the aorta. All the right connections were made for the new heart to perform its function in a new body.

Systems architects often re-use modules across a company's products to reduce cost and design cycle-time while improving reliability. The auto industry is known for leveraging module re-use. Audi re-uses its 4-cylinder and 2.0 liters engine as a module. It appears on the A3, A4, and A5 sedans, as well as on the SUV's Q3 and Q5.

3.1.4 Physical and Functional "Component"

In the physical domain, component assemblies define modules. They are assemblies with reduced complexity to better control their integration into modules. In the functional domain, a component is mostly one or a limited set of functions, which, combined with other functions, generates a stable functional module.

3.1.5 Part and Elementary Function

Items such as a screw, a nut, or a resistor on a printed circuit board perform elementary functions. As such, an elementary function is linked to first principles. Assemblies of parts form components.

Every physical entity has its functional representation. Recognizing this duality by any design team or even by a company is an essential step toward a remarkable ability to innovate.

Fig. 3.2 illustrates the duality of a system's elements in the physical and functional domains.

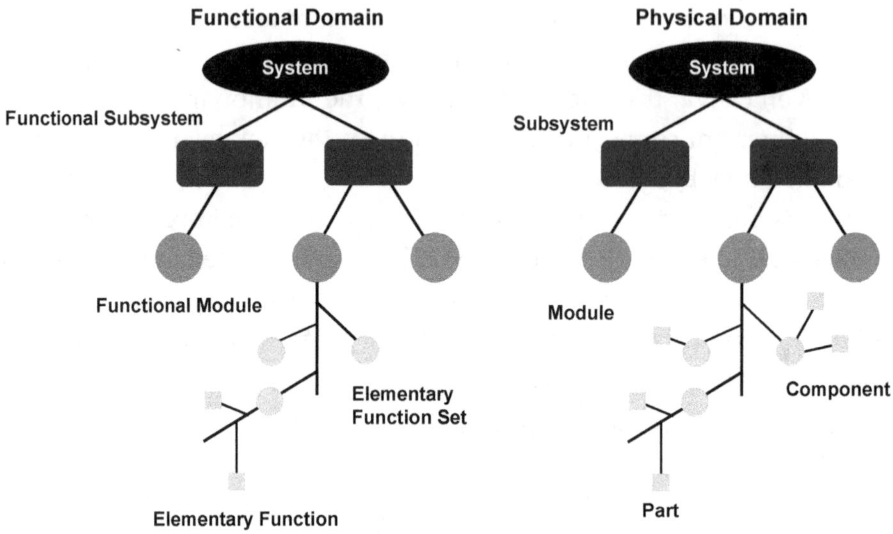

Fig. 3.2 Duality of functional and physical domains

The two figures are not identical due to implementation decisions made during the physical design.

3.2 FUNCTIONAL ARCHITECTURE

Generally, a system's architecture is simply the integrated structure of its elements with their behaviors and relationships. ISO 42010 defines an architecture, in a more formal and widely agreed-upon statement, as "the fundamental concepts or properties of a system in its environment embodied in its elements, relationships, and in the principles of its design and evolution."[1] It is important to note that the broadest definition mentions the elements of the system and their relationships. It is the interactions that allow them to work together to fulfill the system's purpose. In the context of the book, the elements will be the modules considered the building blocks of the systems.

1 ISO 42010: 2011

The functional architecture considers the function sets performed by the modules as functional modules. Being able to identify the distinct function or grouping of functions performed by any system element is critical to understanding what the system does and how it operates.

Going back to the simple example of an empty bottle that performs five functions, the representation in Fig. 3.3A shows the functional and generic physical modules 1, 2, 3, and 4 and their interactions (A, B, and C). The value of functional architecture cannot be emphasized enough. Fig. 3.3A embodies the concept in which the cap, the label, and the handle modules have a relationship with the container. There could be another concept where the label would be on the cap, like in Fig. 3.3B. In that case, the physical synthesis would generate another design. The handle can also be on the cap, like in Fig. 3.3C. So, in general, the functional architecture, when well defined, has the potential to generate other possibilities. New concepts never contemplated before could emerge. Chapter 6 will further expand the notion introduced here.

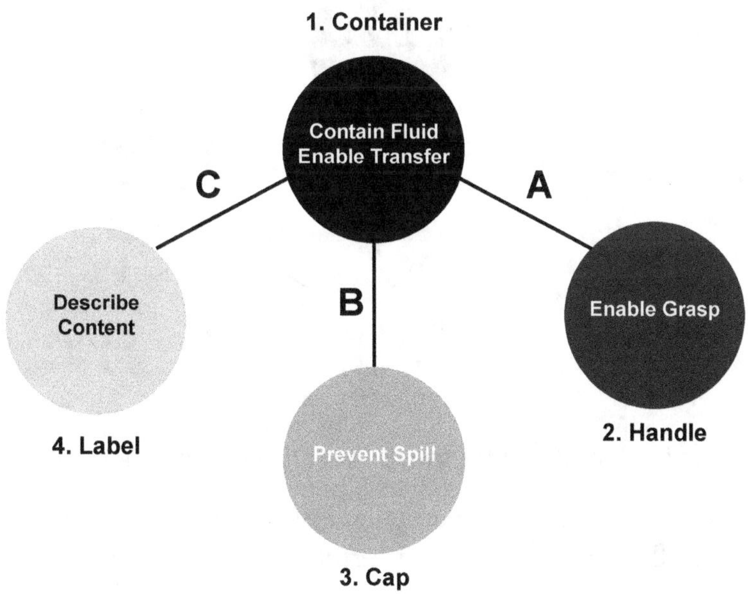

Fig. 3.3A Functional architecture of an empty bottle with its generic modules

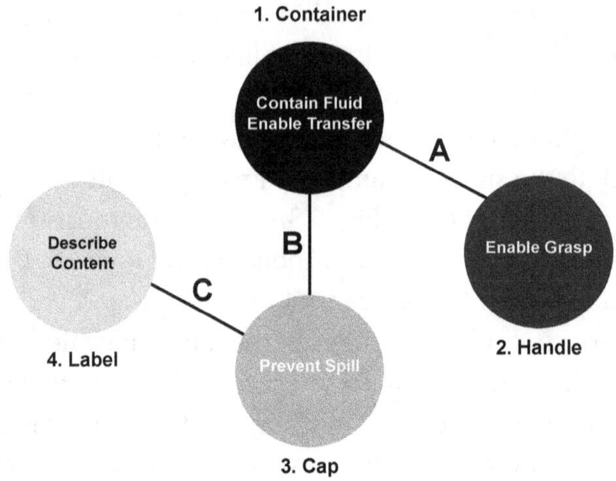

Fig. 3.3B Functional architecture variant of an empty bottle (with the label on the cap)

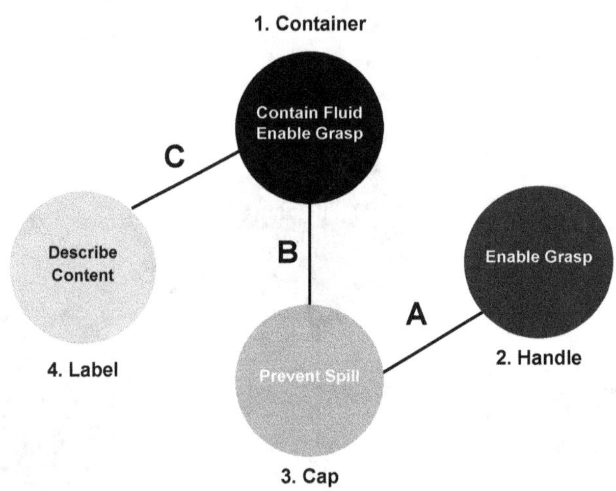

Fig. 3.3C Functional architecture variant of an empty bottle (with the handle on the cap)

Most bottles do not have a handle. The shape of the container "enables grasp." Thus, the container module performs a grouping of three functions, mainly:

- Contain fluid,
- Enable transfer and
- Enable grasp

Therefore, as shown in Fig. 3.3D, in most cases, bottles would be simply represented as below:

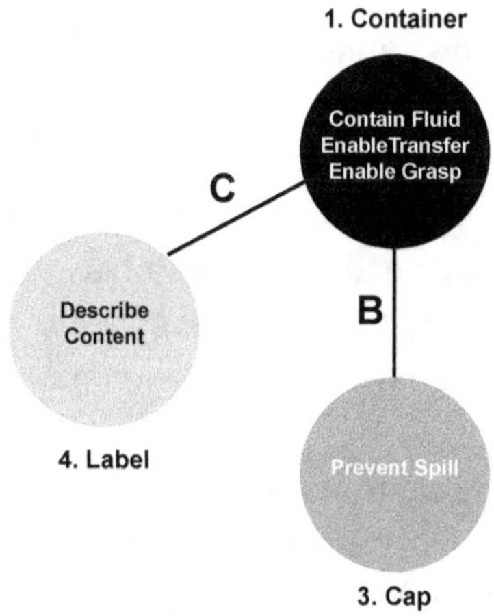

Fig. 3.3D Functional architecture of most empty bottles

Fig. 3.3 highlights how the functional architecture of a system opens new doors to innovation. In the example above, the functional architecture can lead to the design of countless containers for fluids in many phases, liquids or gas, and either hot or cold (See Fig. 3.4).

3.3 PHYSICAL ARCHITECTURE

The transition from the functional architecture to a physical design happens through the system's requirements made of functional and non-functional requirements. Then, the design team takes the implementation decisions

Fig. 3.4 Physical Architecture

and selects the right technologies to meet the needs.

Fig. 3.4 shows various physical embodiments of Figs. 3.3A, B, and C. We can now make a few observations:

- All bottles perform the functions of Fig. 3.3 with different sizes and shapes
- Bottles 1, 2, and 3 are made of glass
- Bottle 4 and 6 are in plastic while bottle 5 is made of treated paper
- Bottles 2 and 3 do not need any labels because their contents can be identified visually by just looking over. But a label would provide more details than just looking inside the container, especially for critical chemical data.
- Bottle 4 has a handle due to its size and shape. A handle performs the "enable grasp" function better than without any in this case.
- Bottle 6 has the handle on the cap. It is the embodiment of Fig. 3.3C.

In general, most bottle designs perform the functions of Fig. 3.3 with only three modules, mainly the container, the cap, and the label. However, there are no limitations to the physical embodiments to create without ever changing the functions. Fig. 3.5 is an example of innovation where a handle

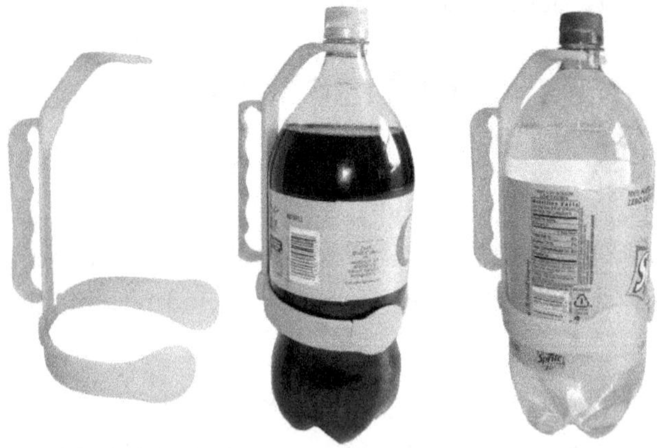

Source: Amazon.com (designed by Cooks Innovations)

Fig. 3.5 Module 2 "Handle" is designed and sold to ease the handling of 2 liters Soda bottles

is commercialized to perform the "enable grasp" function.

We can see how requirements (functions + performance or operational metrics) transform a functional architecture into different physical entities. Understanding a system's requirements from the right interpretation of the identified customer needs is crucial for the success of any engineered system, whether a product, a service, or an organization. Chapter 5 mostly covers requirements structure and content.

Fig. 3.6 below shows how glasses perform only four of the five functions of a bottle. They do not "prevent spill." Yet, they belong to the family of

Fig. 3.6 Physical Architecture: Other Physical Embodiments

containers. Chapter 4 will further clarify the distinction.

The simple principle presented here also applies to much more complex machines once the groupings of their functions are well defined without dependencies. Their modules can be designed with clarity to perform the functions and deliver the expected outputs.

3.4 ARCHITECTURE REPRESENTATIONS

Once the subsystems and their modules are fully defined, the whole system's representation becomes clear. Such a picture highlights the subsystems, the modules, and linkages or relationships that bind them.

As initially shown in Fig. 3.1B or Fig. 3.2, the representation of the functional architecture of a machine, as complex as it may be, is a powerful tool not widely used in all industries. The goal in this section is to draw attention to its usefulness and to promote its use to communicate and make critical decisions. The functional representation of any system starts from the model of the interaction between modules. It is also the expression of their interfaces that could be:

- Mechanical – in such a case, the modules could exchange forces through bolted connections, a fluid, a magnetic field, etc. Even a functional gap, as an intentionally maintained distance between the modules, would be considered a mechanical interface.
- Electrical – electrical power or signals flowing between modules or
- Informational – when a software protocol creates the interaction between modules.

Defining the relationships enables creating the table known as an NxN or N^2 (read N-square) diagram. It is an NxN or N^2 because it is a square with both sides having the same N modules. For convenience we also put their subsystems around them. In each of the intersection boxes between modules that have interfaces, the type of relationship is displayed. The entities could be in contact or even remote, but they still have a relationship of "Input-Output."

The N^2 diagram should be done by the system architects who understand the functional architecture of the whole system.

Let us review a couple of examples to understand the concept.

Example 1: A Bottle Where a Module Performs Each of the Functions.

As stated in the previous sections, the four generic physical modules (functional) of the bottle are:

- The container (functions: "contain fluid", "enable transfer")
- The handle (function: "enable grasp")
- The cap (function: "prevent spill")
- The label (function: "describe content")

Fig. 3.7 Example of a bottle with a container, a handle, a cap, and a label

All of them are physically present in the example represented.

Fig. 3.7 This embodiment represents all four functional or generic modules. Knowing the interaction of the modules, we can complete the N^2 diagram described above.

NxN or N^2	Container	Handle	Cap	Label
Container	1	X	X	X
Handle	X	2	X	X
Cap	X	X	3	X
Label	X	X	X	4

Table 3.1 N^2 diagram of an empty bottle

With the bottle represented, there are no interactions between handle and cap or between label and cap. However, there could be design choices to offer the option to put the label on the cap or the handle on the cap (Fig. 3.4 picture #6), thus creating a link between them. Some companies adopt in their diagram, the designation "X" for standard offerings and "O" for optional. It is simply mentioned here as a notion, but it is the product managers' job to decide what would be part of a standard offering and optional. The idea is presented again in Chapter 11 of the book. In the example, an X is used everywhere for simplicity.

Example 2: Bicycle of Table 2.6

Table 2.6 represents the subsystems and modules of a family of bicycles. It includes motorized and electric bikes as well. Fig. 3.8 is an embodiment of Table 2.6 with the view of a bike with its subsystem and modules. Each subsystem has an icon to distinguish it from the others. Each subsystem's modules have their numbers as [defined in Table 2.6. The representation does not show the subsystems and modules relative to internal combustion engines and electric motors due to the product being a city bike.

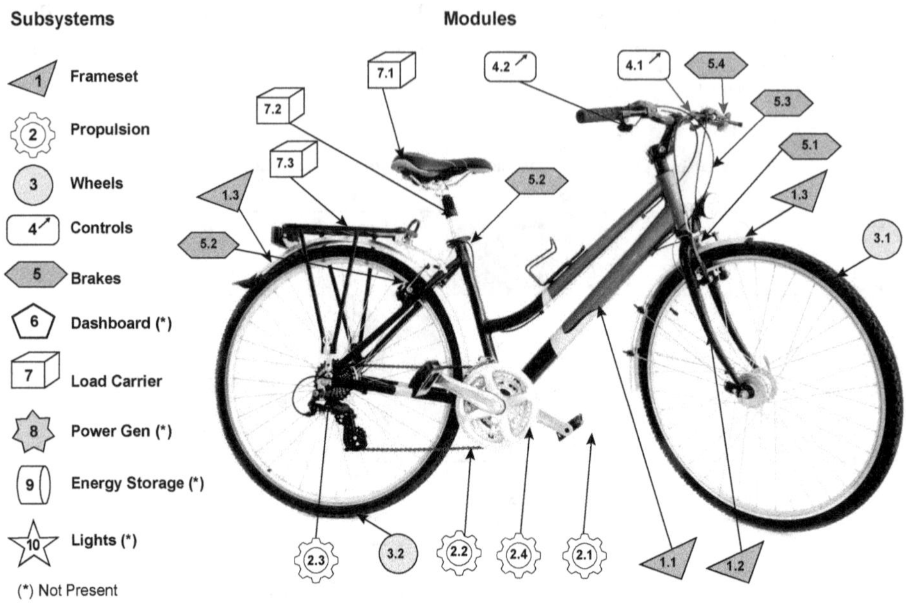

Fig. 3.8 City bike's subsystems and modules

(*) As said above, this bike configuration does not offer the missing subsystems and their modules.

Considering just one subsystem, in this case, Subsystem 1, the representation of the Frameset's modules is in Fig. 3.9.

A module can have one or many components. Module 1.1, the Frame module, is in one piece due to its welded construction. Module 1.3, however, has two components: The front and the rear fenders.

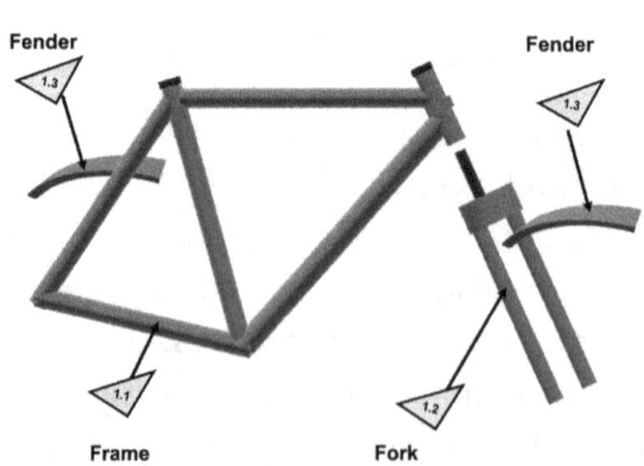

Fig. 3.9 Subsystem 1 "Frameset" with its three modules

Table 2.6 helped create the N² diagram in Fig. 3.10 below with all the modules of a City bike, with an engine powered bike and an electric bike. Each intersection box contains the type of linkage between modules.

				1			2					3	
				Frameset			Propulsion					Wheels	
				1.1.	1.2	1.3	2.1	2.2	2.3	2.4	2.5	3.1	3.2
SS ID	Subsystems	Mod. ID	Modules	Frame	Fork	Fender	Pedals	Motion transmitter	Speed selector	Motion converter	Power converter	Front wheel	Rear wheel
1	Frameset	1.1.	Frame		M	M			M	M			M
		1.2	Fork	M		M						M	
		1.3	Fender	M	M								
2	Propulsion	2.1	Pedals							M			
		2.2	Motion transmitter						M	M			
		2.3	Speed selector	M				M					
		2.4	Motion converter	M			M	M					
		2.5	Power converter										
3	Wheels	3.1	Front wheel		M								
		3.2	Rear wheel	M									
4	Bike Control	4.1	Steering assembly	M	M								
		4.2	Acceleration lever										
		4.3	Stopper	M									
5	Brakes	5.1	Front brake		M								
		5.2	Rear brake	M									
		5.3	Brake cable										
		5.4	Brake levers										
6	Dashboard	6.1	Speedometer										
		6.2	State of Charge indicator										
7	Load Carrier	7.1	Seat										
		7.2	Seatpost	M									
		7.3	Luggage Carrier	M									
8	Power Generation	8.1	Generator	M									
		8.2	Engine	M									
		8.3	Tank	M									
		8.4	Fuel										
		8.5	Electric motor	M									
		9.1	Battery pack	M									

Fig. 3.10 N² diagram section of a family of bikes

ΔP The full diagram is in Section B1.1, Appendix B.

The City bike in Fig. 3.8 can also be functionally represented by Fig. 3.11. It shows the modules and their links or relationships with one another. Modules connected to many others have one-to-many relationships and are "hub modules." Designing of a module, the number of its relationships with other modules determine the complexity of the project. A module connected to only one other module is easier to manage than a hub module. In Fig. 3.11, each node represents a module with its links to other modules. module ws easier to manage than a hub module.

Figures 3.10, 3.11, and 3.12 are all representations of the architecture of the bikes.

Even though all the modules are not offered in one single bike, it is essential to capture the configurations of interest under the functional architecture. Therefore, Fig. 3.12 would have broader use than 3.11. It is the functional architecture of all the configurable bikes considered in the example, including electric and engine-powered bikes. It offers a complete view of all the three types of bikes and opens the way to a better understanding of where the design should go.

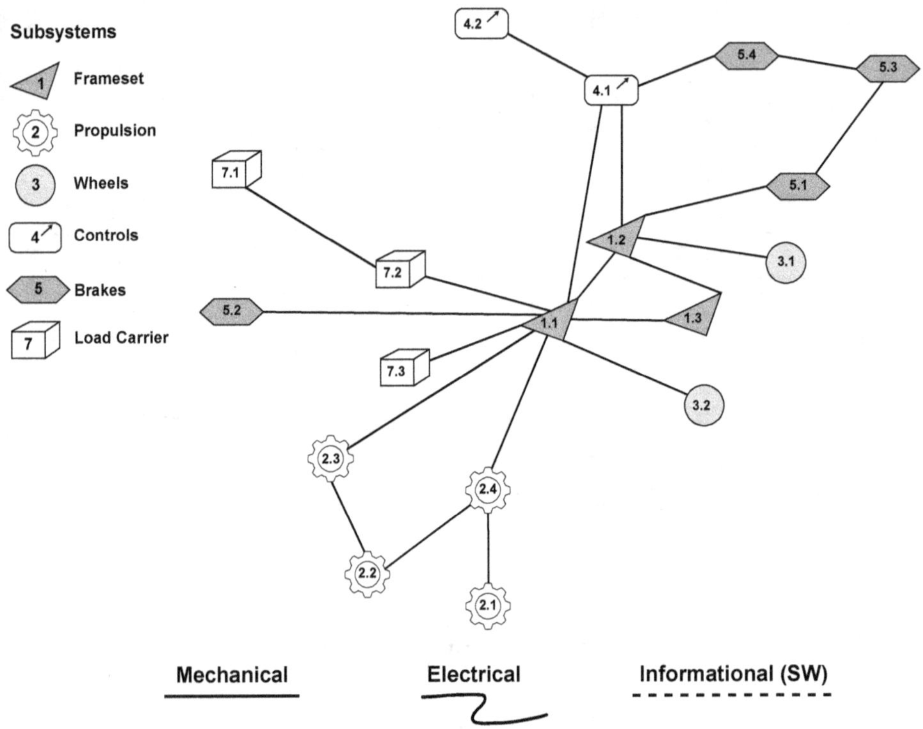

Fig. 3.11 Functional Architecture of a City bike

We can now make a couple of observations about Fig. 3.12.

- Some modules have their interfaces with multiple relationship types, mostly mechanical and electrical or electrical and informational. But it could also be all three.
- Fig. 3.12 is a powerful communication tool between systems engineers and their management. It is easier to point out some modules to show their connections and complexity. When a new project starts, highlighting the new or re-used modules expresses the complexity of the project and the effort needed.

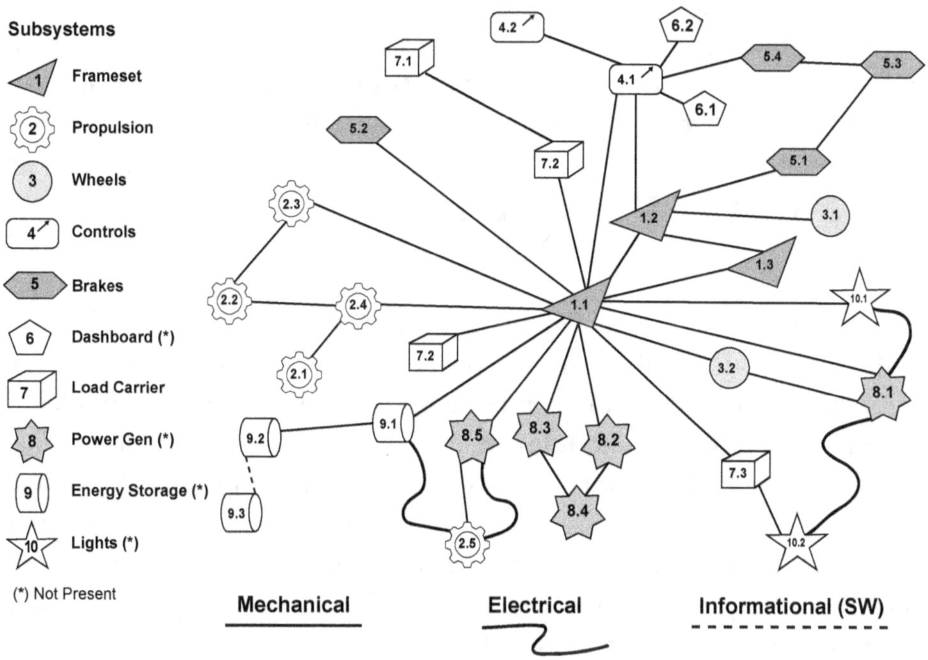

Fig. 3.12 Functional architecture of the configurable bikes from the example

The functional architecture offers a solid base to define the functional requirements, which, combined with the non-functional ones, determine the full design requirements kicking off the design phase. The definition of physical architecture consists of selecting the actual devices that will perform the functions or set of functions. The choices need to comply with all the metrics and constraints imposed by the market. The configurations to offer are also determined. It is when the hub modules must be carefully studied, and a trade-off performed between the minimum number of options of each physical entity needed to cover the whole range and the cost of the complexity of the interfaces. Offering more discrete modules performing the same functions, reduces the complexity of their interfaces that become simpler being specifically designed for a handful of connections. However, this drives the proliferation of modules with high management costs. The alternative is to have multi-functional modules covering multiple configurations with complex interfaces. Therefore, a trade-off needs to be conducted by the system architects and engineers to determine the right balance.

For the bikes of the example, module 1.1, which is the Frame module, is connected to many other modules and requires careful design and manufacturing. Its failure would compromise most of the system's embodiments.

3.5 MODULE INTERFACES

In any system, lower level elements are assembled to form a higher-level element. Considering the hierarchy of system entities, as defined in Fig. 3.1A, parts are developed to build components assembled to create modules. In the context of this book, modules are the building blocks of a system. Therefore, their integration forms the system. Earlier in the book, independence described as one of the main properties of a module enabled its testing alone in isolation from the whole system, based on its inputs to verify its outputs. Such a property allows modules designed in different locations and time and by other teams to be assembled in one place without compromising the integrity of the system. It opens the door for modules re-use across several systems whose sub-elements perform the same functions.

Once a module performs all its functions with the expected performance, meaning, when it meets its requirements, it is ready to be assembled to other modules to form the system. Such assembly is made possible through the module's interfaces to other modules or its environment. The interface of a module, which is how it interacts and communicates with other modules, is so vital that its specification is critical. A module can meet all its requirements resulting in an excellent design. However, if its interfaces are not well defined and designed appropriately, undesired emergent properties or effects can appear at the system's level after assembly and compromise the whole design. Interface requirements must be reviewed and approved by an Interface Control Board (ICB). It is also essential to define them in documents known as Interface Control Documents (ICDs). For a given company, the ICD's define the standard interfaces of the modules that make up their products. Revisiting the example of a bottle, the interface between the cap and the container is a simple mechanical thread. Matching it on the neck of the bottle and inside the cap, the design is open to any bottle shape and size.

Both the bottle and cap could be made anywhere and in different materials and shapes. At the time of their assembly, they will lock tightly into one another and prevent any leakage. The specification of the Interface Control Document, in this case, will simply be the diameter and the thread applied to the cap and the bottle. If beyond leak prevention, the cap must withstand some pressure from the fluid in the bottle, like a soda or high-pressure gas container, the thread must be appropriate, whether it is metric or Whitworth. Several parameters would define the thread:

- Minor diameter
- Major diameter
- Pitch diameter
- Pitch
- Thread angle
- Crest
- Etc.

A module with mechanical links through bolted connections will need the explicit detail of the bolt patterns with all the torques required. It will even detail in which order the bolts need to be tightened. The connections meet the forces they are designed for with the right specifications. Thus, ICD's are requirement documents with all the details necessary for their implementation.

If the links are electrical, the conductors must be sized appropriately based on the operational currents, voltages, and any surge containment.

In the case of an informational link as a software, the communication protocols would be specified.

Popular standard interfaces we all use every day include USB (Universal Serial Bus), an industry-standard that allows the connection of peripheral devices to computers. HDMI (High Definition Multimedia Interface) enables the transfer of high definition audio and video and others.

ICD's will be revisited in chapter 5 with the definition and management of systems requirements.

This section about interfaces may seem very rigid, which is in contrast with a book that is about innovation. To be clear, the management of modules' interfaces is rigorous. Interfaces are so critical that the Change Control Board should be empowered to review and approve changes that will always be possible for the right reasons. Any proposed interface change just needs to be examined to assess its impact on all the systems that use it. It is an effort that is worth doing to avoid costs due to adverse emergent effects. In general, innovation within the modules can be endless, especially when a genuine effort is made to keep the interfaces.

3.6 SYSTEMS ARCHITECTURE SUMMARY

Generally, the term System applies to any entity made of integrated elements forming a unified whole with a purpose not achievable by any of its constituents. "A whole made of integrated elements to achieve a purpose bigger than any of the constituents" is fundamental to a system. Many systems can be integrated to form a system of systems such as a city transportation system of systems made of the traffic lights system, the roads system, the vehicles, etc. In the hierarchy of system constituents, systems are made of subsystems which are made of modules. The modules, in turn, are made of components that are assemblies of parts. Modules are the building blocks of a system. They are used to define the architecture of a system. ISO 42010 establishes an architecture as the fundamental concepts or properties of a system in its environment embodied in its elements, relationships, and in the principles of its design and evolution. It is a widely used definition of an architecture. At the core of any architecture are its elements and their relationships. In the book, the elements considered are the modules. When they are viewed from the lens of what they do, meaning the functions they perform and their relationships, the architecture is a functional architecture. Such a functional architecture allows the system architects and engineers to have a clear view of the possible embodiments or configurations of the product, service, or organization as a human system.

In the case of a product, the functional architecture enables the simplification of its physical offerings, and essential trade-offs become possible to optimize both cost and complexity. The way the functional architecture of a system is defined and represented is described in Section 3.4, and some important observations are made concerning product complexity. Some modules have one-to-many relationships that drive special attention to their development. They are the hub modules, and any change to their design impacts many other modules unless their interfaces are carefully managed. Examples of simple bottles and a family of bicycles illustrated how the modules of the functional architecture or the generic physical modules of a product are defined and represented. A module's interfaces are critical to its design and management for its integration into a system. So, their definition deserved some attention. Well defined and designed interfaces are the key to modules re-use to reduce product development cycle-time and cost, and to improve reliability.

The chapter highlights the value of a product's architecture in providing flexibility to its design and simplifying its complexity. It is a company asset that drives the conception of innovative solutions.

Every company should invest in the creation of its products' architecture. Such an investment will enable the re-use of many proven modules across several product offerings, which will reduce cost, accelerate time to market, and improve reliability.

4

Modular Product Architecture

Chapters 2 and 3 introduced the notions of function, functional analysis, module, and architecture. A module was defined as an independent assembly of components designed to perform a set or grouping of functions. The function sets are formed without dependencies, making the modules that perform them to be independent of one another. So, any module can be tested in isolation from the whole system and have its output validated based on its input. Such a property is one of a module's most essential characteristics. The notion of "interface" was also introduced in Chapter 3 regarding how modules interact or communicate.

All these essential notions introduced in the previous chapters are part of a process that drives module-based product design and management, with an overview shown in Fig. 4.1. The chart highlights the upstream and downstream steps to modules that are the building blocks of a product in its lifecycle. On the left-hand side of the figure, we summarize all the steps described in Chapter 2 and Chapter 3. The right-hand side, which is the subject of this chapter, represents the physical domain and highlights product design implementation and management decisions.

Fig. 4.1 also shows the modules at the center of the functional and the physical domains of the process. A module-based architecture represents a product entirely defined and built from its modules. Such an architecture has a dual nature. It is functional (due to the function sets the modules must perform) and physical (due to the physical embodiments driven by implementation technologies). It is generally characterized as a Modular Product Architecture or MPA. In such an architecture, the module interfaces assume particular importance. Their application enables the integration of modules designed and manufactured anywhere, without compromising the integrity of the whole system. Modular product design is the key to mass customization.

This chapter will also describe how a modular product architecture drives excellent simplification of a product, its configuration, its assembly, its service, and even its management at the end of its First Use.[1]

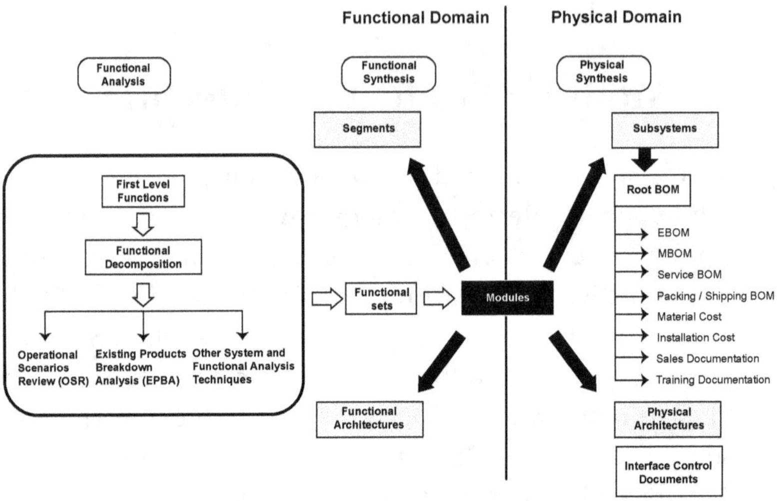

Fig. 4.1 Modular Product Architecture Process Overview

As shown in Fig. 4.1, function sets (function groupings) determine the functional architecture and uniquely define the generic physical modules. In Chapter 3, Fig. 3.3A shows the generic modules of a bottle as a container, handle, cap, and label.

Once the functional modules, which are also physical and generic, are defined, the transition to the physical domain takes shape through the implementation decisions about which technologies to apply to build and market. For example, the container of a bottle is functional as it performs a set of functions but also generic as it represents a generic category of devices that contain fluids or loose particles. A user would grasp it to transfer its content. It becomes physical through the synthesis or design process, including the

1 We prefer the term "end of first use" to the commonly adopted terminology "end of life." In the 21st century, we should not be designing products that have anything thrown into a landfill at the end of use. Their components or materials should be re-cycled to give birth to new products or the product itself should be re-moduled in an upgraded architecture or even a new architecture.

choice of technologies and procedures to build it. Fig. 4.2 shows an example of the transition of a container module from a functional to a physical state. Its design defined shape and material.

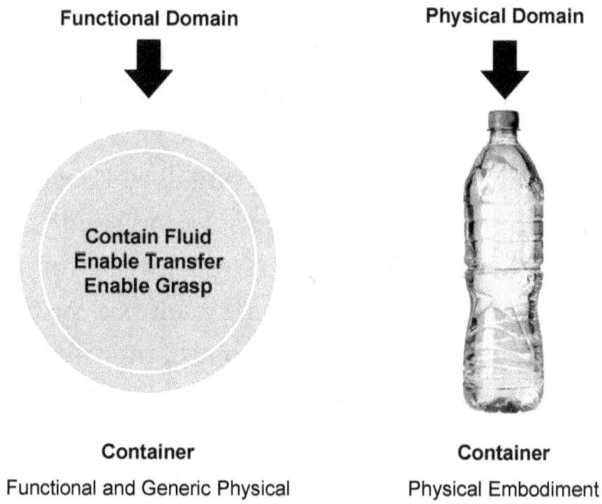

Fig. 4.2 Container module in functional and physical domains

Most people may be familiar with the concepts presented in the next sections of the chapter as we explore the physical domain of the process described and represented by Fig. 4.1. We will also review the notions of instances and variants, Core and non-core modules, and the definition and use of platforms. In the end, the modular product architecture (MPA) is a means of achieving a product that satisfies customer needs. The actual product delivery requires essential information and documentation that captures the Bill of Materials (BOM), material and installation cost, sales information, etc. The chapter also covers how these can benefit from a modular product architecture.

4.1 INSTANCES AND VARIANTS

Before diving into modular products, it is crucial to revisit the concept of a system. In Chapter 3, the term system was, in a broad sense, applied to any entity made of integrated elements forming a unified whole with a purpose not achievable by any of its constituents. We also defined a system of systems and a subsystem. Among the systems architecture principles described in

Appendix E at the end of the book, Principle #2 states that "every system is the subsystem of a larger system; and for Principle #3, "Every system of systems expands towards the universe." Thus, every time a discussion starts about a system, its boundaries need to be specified. Its operating environment also needs to be clarified to avoid wasting the time of those involved as they talk about two different entities. A quick exercise will help understand important concepts later in the book.

Let us assume for a minute that I am sitting at my desk in my office. My desk is a system made of its working area, its cabinets with drawers, handles and keys, its outlets, etc. It was designed, manufactured, assembled, and delivered to be part of the office. The office itself is a system that includes the desk described, my computer, my bookcase, my meeting table, and all the chairs. I interact with the office system every day to perform my professional duties. Now, my office is in one of the buildings which are part of the campus. The campus itself is part of the city. We can place the city in the county, then in the state, the country, the continent, planet earth, our solar system, our galaxy, and our universe. This quick exercise matters when we define the context in which we talk about instances and variants. To avoid any confusion, going forward,

- We will apply the term System to a specific integrated assembly with clear boundaries and operating environment. We will call it our system of interest (SOI). Therefore, the system of interest needs to be defined to allow the conversation to be meaningful. We cannot be discussing an office when other members are arguing about the building or the campus and vice versa.
- Instances, in the context of the book, will be mainly related to modules and sometimes to components but not to subsystems or systems. The Merriam-Webster dictionary defines an instance as an example, a case, a specimen that exhibits distinguishing characteristics in its category. So, instances characterize a category of modules. We could have an instance of any constituent of the system's hierarchy, but the choice made here is to reserve the term instance to modules and components. Describing examples of systems, subsystems, or even system of systems, we will use the terms system configurations and system offerings or only configurations or offerings referring to what is offered by the companies

and can be purchased.

- Variants distinguish modules that belong to the same category represented by an instance. The term will be used only with modules and sometimes with components.

Without such a convention for this book, everyone would not see the same thing at the same time. Here are a couple of examples.

Example 1: Bottle Offerings

Fig. 3.3A identified the four physical generic modules of a bottle as:

- The container
- The handle
- The cap and
- The label

The observations made in Section 3.3, supported by Fig. 3.4, showed that some of the container modules of the bottles were made of glass, others in plastic or treated paper. Some of them have the labels printed on them and some not. Some include a handle, while most of the others have a shape that allows grasp. In terms of instances and variants, each instance of a container of a bottle can have several variants where the instance, as said earlier, represents a specific category. Generally, once an instance is defined, variants are created by varying some attributes, properties, or features that can respond to functional and non-functional requirements. Fig. 4.3 represents two instances of container modules.

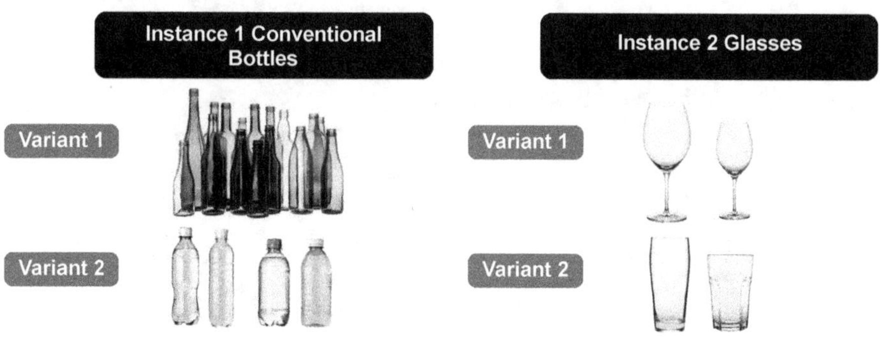

Fig. 4.3 Examples of container module instances and variants

Some readers may be surprised to see glasses considered as instances of a container module. Thinking in terms of functions opens our minds to new possibilities. In this case, we know the instances of the example. In general, new instances that were never previously considered emerge. We will revisit the notion in Chapter 6.

We should note that size does not differentiate variants, as shown in Fig. 4.3 above.

Example 2: Bike Offerings

Let us consider module 1.1 for bike offerings, which is the frame. Fig. 4.4 below shows two instances of frame modules of city bikes with a few of their variants. The functional or generic physical module is "frame." Instance 1 is "men's bike frame," and instance 2 is "women's bike frame." As stated in the previous example, the variants should not be confused with the frame sizes. Each of the variants still comes in various sizes based on the rider's height. Each instance and its variants are decided based on market requirements. For this example, the variant drivers for each instance could be a shape (as shown) or material (steel, aluminum, carbon fiber, etc.) or other requirements.

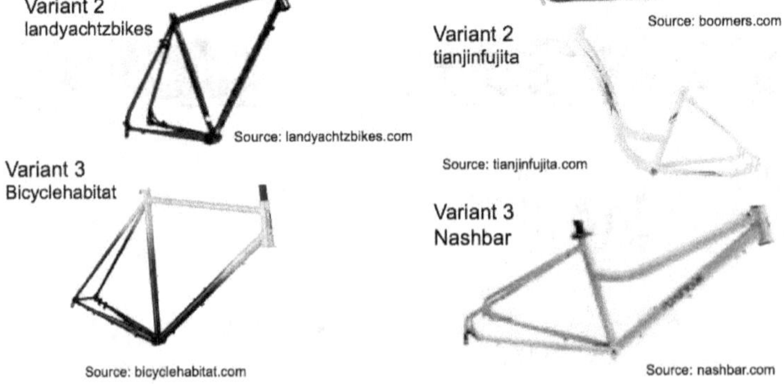

Fig. 4.4 Examples of City bikes Frame module instances and variants

Good product design starts from an architecture that establishes the modules and their relationships. Then the module instances and their variants are defined based on market needs. Offering a series of variants to address several markets from a limited number of generic modules is vital. The winning solution is the one where the instances are also limited but can give birth to the required variants.

Let us consider Table 4.1 below, which defines the Portfolio Synoptics Matrix (PSM) of bike offerings. A PSM is a company's offerings overview with the generic physical modules. It helps understand how a company's modules map to the offerings before considering module instances and variants.

				Bike Offerings					
SS ID	**Subsystems**	**Mod. ID**	**Modules**	City bikes	Mountain bikes	Race bikes	Tandem bikes	Motorized bikes	Electric bikes
1	Frameset	1.1.	Frame	X	X	X	X	X	X
		1.2	Fork	X	X	X	X	X	X
		1.3	Fender	X				X	X
2	Propulsion	2.1	Pedals	X	X	X	X	X	X
		2.2	Motion transmission	X	X	X	X	X	X
		2.3	Speed selector	O	X	X	O	X	X
		2.4	Motion converter	O	X	X		X	X
		2.5	Power converter						X
3	Wheels	3.1	Front wheel	X	X	X	X	X	X
		3.2	Rear wheel	X	X	X	X	X	X
4	Bike Control	4.1	Steering assembly	X	X	X	X	X	X
		4.2	Acceleration lever	O	X	X		X	X
		4.3	Stopper	X	X	X	X	X	X
5	Brakes Assembly	5.1	Front brake	X	X	X	X	X	X
		5.2	Rear brake	X	X	X	X	X	X
		5.3	Brake cable	X	X	X	X	X	X
		5.4	Brake levers	X	X	X	X	X	X
6	Dashboard	6.1	Speedometer	O	O	X	O	X	X
		6.2	Charge indicator						X
7	Load Carrier	7.1	Seat	X	X	X	X	X	X
		7.2	Seatpost	X	X	X	X	X	X
		7.3	Luggage Carrier	O					
8	Power Generation	8.1	Generator	O	O				
		8.2	Engine					X	
		8.3	Tank					X	
		8.4	Fuel					X	
		8.5	Electric motor						
9	Energy Storage	9.1	Battery pack						X
		9.2	Charge supervisor (SW)						X
10	Lights	10.1	Headlight	O			O	X	X
		10.2	Rear light	O			O	X	X

For example, Table 4.1 shows a few bike offerings and how the generic modules map to them. It describes how a limited number of functional

Table 4.1 Portfolio Synoptic Matrix (PSM) of Bike Offerings. "X" and "O" represent standard and optional offerings.

modules cover a wide range of bikes. Even though each bike offering may have a different module instance, their comparison becomes possible through their subsystems and modules. Although not very explicit, behind the PSM, each module instance for any specific bike offering may have several variants, as shown in Fig. 4.4. Thus, there are several configurations or offerings of bikes, namely, city bikes, mountain bikes, race bikes, etc. Specific module instances with their variants characterize each configuration.

In the bicycle world, the variant of men's bike frame (module 1.1) made in aluminum or carbon fiber to be lighter fits race bikes, and the solution is prevalent.

ΔP Another example of PSM is in Section A.1.3 of Appendix A in the supplement to Chapter 2 (Table 2.8 for a Vapor Compression Architecture).

4.2 TAXONOMY

With the notions of subsystem, module, module instances, and their variants introduced, a company is ready to set the taxonomy of its products.
Armed with the functional architectures of their existing products, companies can set the best classification and language that will bring all the various departments, regions, divisions, and business units together. To stay with the same examples, a bike company has ten subsystems and 32 functional or generic physical modules. Independently of the bikes' configurations on the market, it is essential to adopt the subsystems with their names. Subsystems and their generic modules are derived from the functions they perform to drive customer value by a company, regardless of the technologies used to make them. Therefore, their designation is stable over time.

Based on such a convention, the bike company can assume Subsystem 1 always to be the Frameset, 2 will be the Propulsion, etc. The same way at the modules level shows that:

- Module 1.1 will always be the frame
- Module 1.2 will be the Fork
- Module 1.3 will be a Fender (bikes sold without fenders will not have module 1.3 offered)

Following the convention, the designation of each module would have the first number linked to its subsystem and the second one its order within the subsystem. So, module 1.1 is the first module of the first subsystem. When a company adopts such a language, people do not even call the name of the modules anymore. It is always amazing to hear a high-level executive ask in a meeting how many variants of the first instance of a given module are in consideration and why? The rest of the company understands immediately. Comparison between module instances and even their variants become regular discussions. The business unit that offers city bikes and the one that specializes in race bikes can still discuss using the same language and compare cost and features. A genuine question like: "Why can't we use the same instances of module 3.1 and 3.2 for a city bike and a motorized bike?" would be familiar.

The classification of company product portfolios into subsystems and functional modules followed by the module instances and variants has advantages that the practitioners and executives can understand and appreciate.

Companies evolve organically or by mergers and acquisitions. In mergers or acquisitions, the companies come with their own cultures and their product portfolio. Some of their products will overlap, but some will supplement each other in terms of market segments and offerings. A post-acquisition activity is to align the new company's products to the existing taxonomy of the host company. It quickly highlights the golden nuggets acquired and how their leverage can reduce the cost of the current products or provide more offerings from a limited set of module variants.

Another critical application we will review in Part II of the book is about the supply chain. Considering the supply chain, the company could have Module Integrators. They would provide the Modules by manufacturing them based on an internal design or even design them to meet a set of internal requirements. In that case, the strategic supply partners will strictly respect the interface requirements included in the ICD's (Interface Control Documents).

Some companies go even farther to extend the supply partnership to Subsystem Supply Partners (SSP's). Each SSP could own a whole subsystem from design to manufacturing and deliver it for full system integration.

Chapter 7 in Part II covers the application of a Modular Product Architecture to the supply chain.

4.3 MAPPING MODULES TO INSTANCES AND VARIANTS

It is crucial to map the module instances and their variants of the company's product portfolio to the taxonomy defined in section 4.2. It is a laborious and time-consuming exercise.

The mapping exercise forces the systems engineers to revisit all the module instances and their variation drivers. Some drivers are presented below as examples to help understand the complexity of the kind of work required. The following sources could drive variation among the same instance (same module, same instance but different variants).

- Regions or Countries
  - Global Codes and Standards (# Voltages, #Sizes, etc.)
  - Local regulations
  - Known cultural preferences
- Market Segments
  - Types of offerings (Number of applications types)
  - Product Range (Low-End, Mid-Range, High-End)
- Company differentiation choices
  - Features (Number of options)
- Etc.

It appears quickly that the company is maintaining an incredibly high number of variants.

The illustrative Fig. 4.5 gives a sense of the number of variants that could be in play if each of the arrows from top-down, indicating a system element, carries the number of items below it.

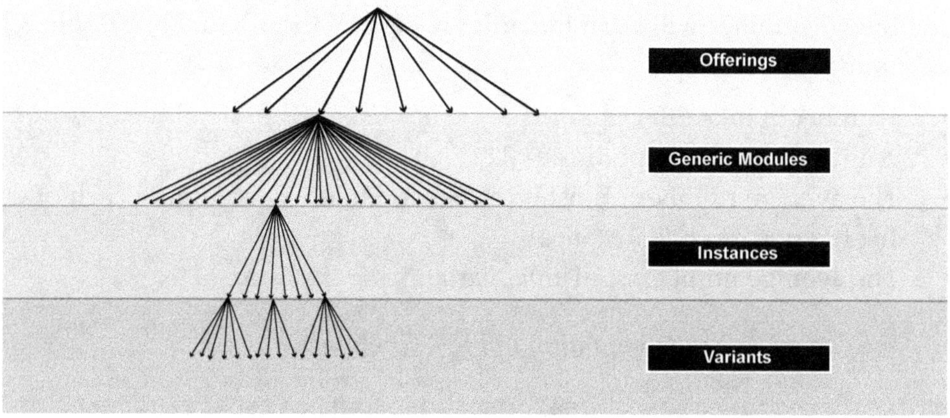

Fig. 4.5 Product Offerings, Modules, Module Instances, and Variants

Each offering has a set number of configurations made of subsystems broken into modules with their instances and variants.

To give a sense of how quickly complex a product portfolio can become, let us review some of the leading independent variables. Given,

- Number of offerings – n
- Number of generic modules – m
- The average number of instances per module – p
- The average number of variants per instance – v

To explain the complexity, we would assume that for each offering, only one distinct variant would count out of quantity v variants. Even if the total quantity picked is more than one, the assumption would still be that they all have the same part number. Only one instance out of quantity p instances would part of the offering. With these assumptions, the number of modules per offering would be:

- Number of module variants per offering = C(1,v) x C(1,p) x C(1,m) = v * p * m

Therefore, the total number of module variants to be designed to cover all the offerings would be:

- Number of total distinct modules variants = v * p * m * n

Such a number will be in the order of several thousand. From Table 4.1, assuming:

- Number of bike offerings – 6
- Number of generic modules – 32
- The average number of instances per module – 6 (assuming each offering has one module instance)
- The average number of distinct variants per instance – 4

We can get the representation in Fig 4.6 below:

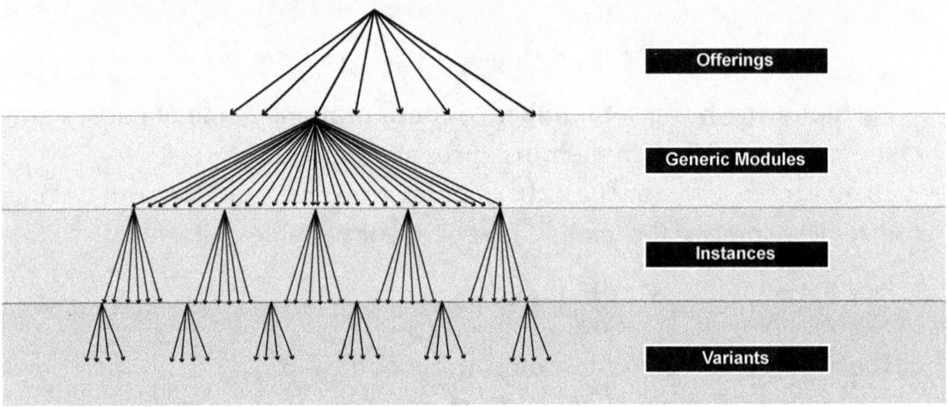

Fig. 4.6 Bike Offerings, Functional Modules, Module Instances, and Variants

Even though we know that not every offering gets the same number of functional modules, we assumed for simplicity that they all did. So, the number of module variants per offering would be 768, and the total number of module variants for the whole portfolio would be 4,608. Such a result is for only six types of bikes. Beyond all the simplifications we made, companies offer much more system embodiments with configurations that drive more variants than the given example. For that reason, when companies perform the mapping of their existing product to the newly defined subsystems and functional modules, they are shocked by the large number of variants they are maintaining with huge implications on inventory, product cost, lead time, contribution margin, cashflow, and more.

Once the mapping is complete, the company's Product Management and Marketing leadership, with the support of Engineering, Supply Chain, and Service need to revisit and define:

- The market segments to address,
- How to address them (Product portfolio, single offerings, etc.) and,
- When to launch any new defined offerings

The data will lead to the alignment of requirements and help set the number of offerings, configurations, module instances, and their variants and, most notably, a product roadmap. Once completed, then new systems design can get underway. The existing module instances and their variants are mapped to the requirements to determine which ones best meet them across multiple configurations and offerings. After their selection, the team can fill the New, Modify, and Re-use table template below:

Subsystems	Functtional Modules	Module Instances	Module Variants	Re-Use	Modify	New

Table 4.2 New, Modified and Re-use Matrix

The complexity of any program becomes apparent, and its work packages can be planned and executed. One of the project outcomes is a greatly simplified Bill of Materials (BOM) and reduction of other essential, but hidden complexity and variation management cost.

The process can go one step further to revisit all the modules of the company's product portfolio targeting entire subsystems or groups of modules from different subsystems re-used across a whole set of offerings from other markets. Such arrangements are known as platforms, and Section 4.5 will cover their detail.

The process and examples emphasize the re-engineering and re-organizing existing products or systems because pure New Product Development (NPD), where all modules are new, are relatively rare or represent a small percentage of the products developed by companies around the world. Even with NPD, the company still needs to define the markets to address, the configurations to offer, and the Product Synoptics Matrix (PSM). For both NPD and existing upgrades, after setting the PSM, the module instances and their variants are defined to minimize their number by leveraging their re-use. While we have not yet discussed cost, due to the complexity of the topic, it is still worth noting that during the project, the whole system cost optimization should take priority over the optimization of a single variant cost.

The most crucial objective of all the initiatives is to define the minimum set of module variants that will cover the most extensive product configurations possible. The goal should be the same either with existing products or new first-generation products.

World-class companies can re-use 80% of their modules from one product generation to another.

The next section shares a couple of examples to emphasize the power of re-use.

Example 1

At the Ingersoll Rand Power Tools business unit, the evolution from a Pistol (vertical) Impact Power Tool to an Angle (Horizontal) version took only six months instead of the usual 18 months. There was an engaging discussion with the team. Despite all the explanations provided, the engineering leader did not compromise on the opportunity to re-use all the electronics, the battery pack, the controls, and other essential modules. The team accepted the challenge and achieved re-using 89% of the modules. Despite the new shape, the Angle Tool's internal modules were all the same as the Pistol one.

This innovative new design convinced the few skeptics, and the same team later achieved several other significant applications of modules re-use.

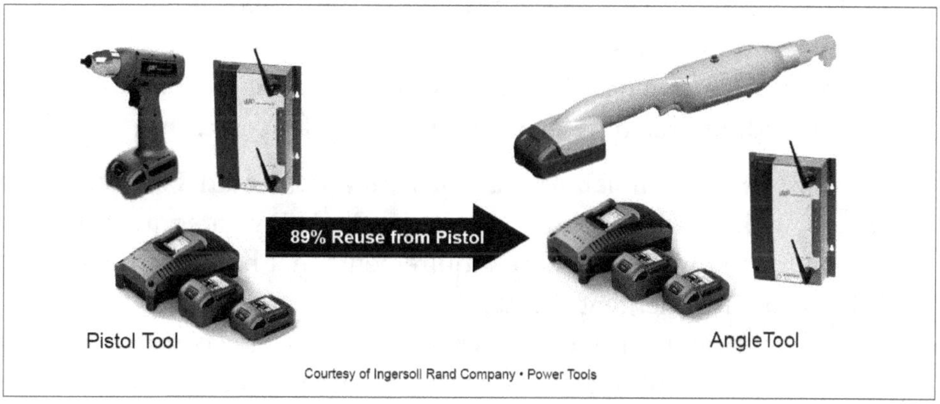

Fig. 4.7 Conversion of a Vertical Tool to a Horizontal Tool

Example 2

The design team at Ingersoll Rand Company – Power Tools Business Unit could leverage existing air operated tools to rapidly design and release new electric cordless tools to the market, as shown in the figure below.

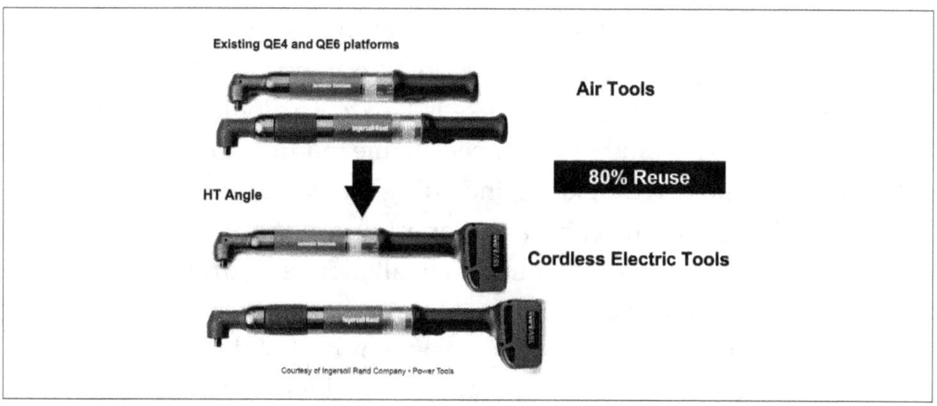

Fig. 4.8 Conversion of an air-operated tool into an electric cordless tool

Once a development team gets it, they substantially reduce product development cycle-time and continue to improve the design process. Companies should invest in leveraging modular product architecture to:

- Increase offerings from a limited set of modules,
- Gain full control of the number of module instances and variants to offer,
- Reduce product development cycle-time,
- Reduce the total cost of the product overall, even if some module instances and variants may see some increase,
- Improve product reliability by maximizing the re-use of existing and vetted modules that have been on the market for some time.

Improve product reliability as each module would be tested independently with clear input and output before being integrated into the system by following the interface requirements that have been defined in the ICD's to avoid emergent defects.

Even though all the modules of a system are essential constituents of its design, delivery and ultimately of its success, it helps companies adopting a modular product architecture to classify the generic physical modules into Core and Non-Core modules. The next section explains the implications of such a classification.

4.4 CORE AND NON-CORE MODULES

One more critical notion to introduce that Part II will revisit is the notion of Core and Non-Core modules of a system. Products or, in general, systems offered on several global markets require important investments from the companies. Therefore, module re-use is key to accelerating development and reducing the investments they need. Some modules are driven by some critical technologies to companies, including but not limited to trade secrets. Such technologies set the companies apart and differentiate them on the market. Their design, manufacturing, installation, etc., drive a competitive advantage. Their re-use crosses different market segments and portfolios. Modules or groups of modules that meet all or part of these criteria should be labeled "Core" to the company and be treated with all the attention they deserve.

The categorization of modules as Core comes with several implications:

- Core modules should be preferably designed internally to preserve their unique attributes and protect the investments made for their development.

- Core module manufacturing should be internal or by Subsystem Supply Partners only. Such a choice would also protect the competitive advantage and investments.
- All core modules' interfaces must be controlled and have an ICD. Some companies may decide, due to cost, not to handle all the module interfaces of their systems. But an additional effort should be made to expand the control to the interfaces of other modules connected to the core modules.

Non-core modules are also necessary but not critical to the company's competitive differentiation. If they are part of a platform (which is in the next section) designed for global application, their interfaces should be treated like core modules and controlled through ICD's.

Another critical step in support of achieving some of the benefits of an excellent modular architecture mentioned above is to define what is known as platforms.

4.5 PRODUCT PLATFORMS

A product platform is a defined set of subsystems or modules whose applications go across the offerings of a portfolio. The re-use of some defined elements drives the coverage of large variations of offerings. Any platform design follows an assessment of the functions or the features to be leveraged across the portfolio. Although it may be an expensive investment at the start, it helps the company save enormously through time to market reduction of new product design, manufacturing, testing, and release. The automobile industry makes heavy use of product platforms. The industry uses this strategy on different types of cars (sedans, crossovers, or sport utility vehicles) to share the chassis, engines, and more from a hardware standpoint and even more from the software perspective.

In other heavy machinery industries, human-machine interface (HMI) controls (including both hardware and software), and some basic structures are mostly leveraged through platforms. A platform design requires good functional and physical architectures. In a modular architecture environment, the subsystems and modules are conceived with an initial re-usability in mind. It is said to be initial because it usually evolves as new products are added to the portfolio. A platform design tends to aim at optimization at a higher level than the single component or module. In our case, we will consider a modular product architecture environment.

Fig. 4.9A and B show how a platform can be designed by leveraging a whole subsystem or a group of modules.

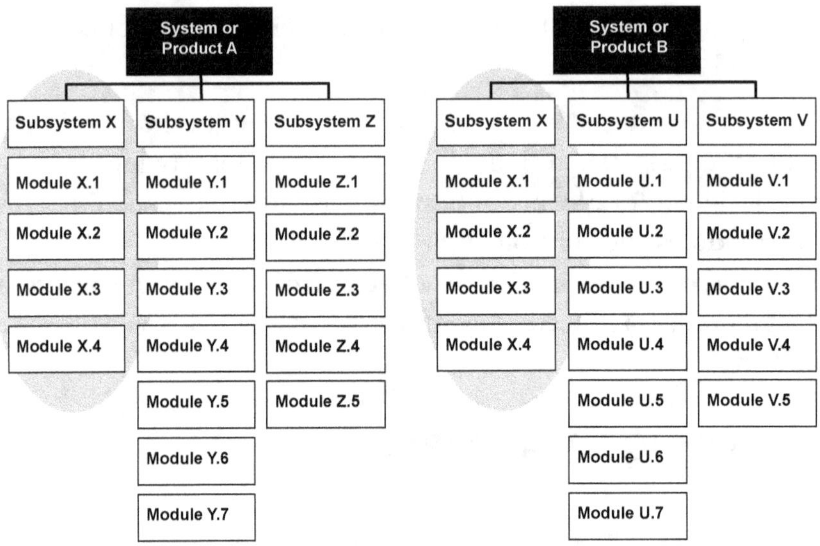

Fig. 4.9A Subsystem Leveraged as a Platform

In Fig. 4.9A of subsystem X with all its modules, X1 through X4 are re-used across a whole range of products to save money while reducing technical risks and development cycle-time. Both products A and B use all the modules of subsystem X. Such a subsystem could be mechanical, electrical, a combination, or even software. Once adopted, the other subsystems not part of the platform present the opportunity to differentiate the offerings further. Since companies conceive platforms to cover several product offerings, they need to pay special attention to low-end, mid-range, and high-end markets. There is a customer engagement risk when the offerings are not differentiated enough; high-end customers could stop buying as their choice could be the same as the low-end with the perception of being cheap. Some car companies have paid the price of reduced sales in the marketplace by not differentiating enough in their vehicles' features.

In Fig. 4.9B, a set of modules that are re-used across several products define the platform. Such modules also come from different subsystems. In this case, modules X.2 and X.4 from subsystem X, module Y.1 from subsystem Y, and module Z.5 from subsystem Z are re-used in both products A and B.

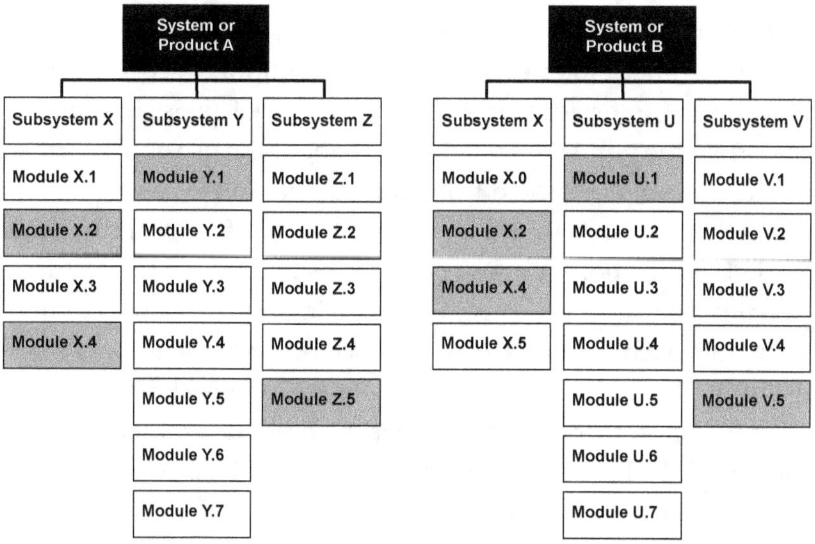

Fig. 4.9B Group of Modules Leveraged as a Platform

The company could use the same modules in other portfolio products beyond the system or product A & B. Platforms can also be the combination of entire subsystems or even a mix of modules from complete subsystems and other discrete modules across other subsystems. Several factor such as standardization to drive re-use, cost and quality, dictate the choice of the modules that make up a platform. Still, it can be made simpler by leveraging the set of functions performed across a portfolio's products or even across portfolios. All this is facilitated by the interface control documents (ICD's) mentioned in Section 3.4 and reviewed again in Chapter 5.

Once established, platforms enable a quick global application of the product offerings involved. They also help reduce manufacturing and supply chain investments as the technologies are replicated with all the lessons learned. Some detractors of the platform strategy often raise the risk of potential multiplication of issues and a more significant impact in a product recall case. They argue that platforms cover a wide range of offerings, and in case of a defect on mass-produced products, the numbers can be staggering with heavy financial losses. Considering the Power Tool example in Fig. 4.7, if the electronic modules inside the tools, the battery, and the controller were part of a platform, a defect to the batteries would have a devastating impact on all the tools that leverage them. The detractors' case should not be simply dismissed. Their arguments call for more rigorous attention to the design and verification of all the modules of a platform. The definition of their interfaces with other modules also deserves additional scrutiny. Therefore, the importance of Interface Control Documents (ICD's) is significant.

To illustrate the point generically made about the use of platforms, let us observe a few examples picked from general literature.

In 2019, Honda announced the introduction of "Honda Architecture" with market application in a few years. It is a flexible architecture based on a global platform that will reduce vehicle development cost and time. Their new approach is not surprising since other car manufacturers have been using platforms. Toyota is known to base several cars on its Toyota New Global Architecture or TNGA that inspired several platforms. After its launch in 2015, Toyota has leveraged the architecture to cover several platforms and categories of vehicles:

- Subcompact vehicles use TNGA-B. This platform covers the Toyota Yaris.
- Compact vehicles use TNGA-C which covers Toyota Prius, Toyota C-HR, Toyota Corolla, and Lexus UX, among other cars.
- Larger sedans use TNGA-K, and it covers vehicles such as Toyota Camry, Toyota Avalon, Toyota Highlander, Toyota RAV4, and even Lexus ES.
- Luxury sedans use TNGA-L that covers vehicles such as Toyota Crown and Lexus LS.

These platforms cover even more cars than listed above, which are all derived from the New Architecture. They use similar unibody structures and other components the vehicle manufacturer called "Core Components" designed to accommodate different vehicle types while differentiating them from one another with non-core features. The platforms were not limited to new body type, new suspensions, and safety features; they also extended to new manufacturing methods and new materials. As shown from the list above, they cover sedans and sport utility vehicles (SUVs).

Design is not complete until the product is manufacturable, installable, serviceable, and recyclable. In that regard, one of the crucial documents of the process is the Bill of Materials (BOM), which is described next in an MPA environment.

4.6 BILL OF MATERIALS

The level of detail in this section is intentional due to its importance. It is in the main body of the book rather than the Appendix

As stated above and reinforced here, in premier companies, the design is not complete until it includes all the processes relative to the product's manufacturing, sourcing, assembly, shipping, commissioning, service, and recycling. Teams from marketing, engineering, manufacturing, supply chain, field installation, and field service, supported by human resources and finance, should make all these decisions during the product's development. Once the design is complete, among the critical documents released is the Bill of Materials (BOM). It represents the transition from digital models and drawings to the data details of all the items (subsystems, modules with their components, parts, and raw materials) and their quantities needed to configure a product, manufacture it, assemble it and deliver it to customers. The BOM is the bridge between design engineers and manufacturing engineers. The BOM list breakdown occurs at all different levels of the product. Some companies call these levels:

1. The unit level
2. The options group
3. Sellable and make-able options
4. Components, sub-assemblies, and parts

Some companies have two types of BOM. The engineering bill of materials or EBOM and the manufacturing bill of materials, MBOM. Companies that design and manufacture products in multiple locations often adopt such a distinction. When design engineers finish their work, they create the Engineering Bill of Materials or EBOM, which is the structure of all levels of the product as designed, from the system to the modules, components, and parts. Once the EBOM is complete, manufacturing engineers take it and create the Manufacturing Bill of Materials by adding more detailed information necessary for production and supply chain operations. The proponents of EBOM and MBOM claim more flexibility in decoupling engineering functions from manufacturing and sourcing. But complications occur when EBOM and MBOM are not tightly linked and begin to drift from one another. Even with all the best intentions, a new product released with EBOM and MBOM in sync sees their divergence after a few inevitable revisions for very legitimate reasons. Over time, the two documents can evolve into different versions with modifications that are not in phase, becoming the root cause of some product quality issues. The problem is seen when experienced factory workers make parts that are not to specs but still work due to their experience. They usually know the changes' history and can make the pieces work despite not meeting the drawing requirements.

A modular product architecture offers some opportunities. It defines a product structure with the subsystems and the modules that become standard for each company, as described in Section 4.2 about taxonomy. Such an arrangement is relatively stable and is common to all the products of the same portfolio. It can also be defined to cover several product portfolios.

A product's data structure, including its modules, their instances, variants, and components with their materials and quantities, defines the Root Bill of Materials (RBOM). Engineering delivers the RBOM as an out of the design work. The document should be common to all the products of the same portfolio. After its completion, manufacturing engineers add the data needed for production and sourcing to create the Bill of Materials or BOM. It should be one document managed by both engineering and manufacturing and be under revision control.

The RBOM can assume any form a company decides to adopt. Its formal presentation is not as critical as its information content. Whether derived directly from a 3D model or tables, Fig. 4.10 summarizes the information it needs to contain.

Fig. 4.10 illustrates the elements of the RBOM. The shaded levels 1, 2, and 3 should be standard for any company. They represent the functional architecture, as seen in the example of Table 2.5 for a portfolio of bikes.

Engineers who design the products should assign "Root Numbers" to each system, its subsystems, and modules. Products of the same portfolio should be distinguished by either a prefix or a suffix added to the generic root number. Some companies define the naming convention that simplifies their database and helps them find any existing parts easily without recreating

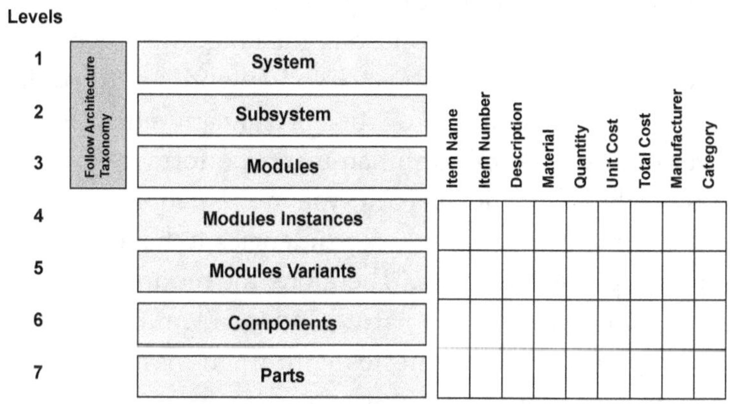

Fig. 4.10 BOM information content

them every time not finding them. Other companies leverage the Computer Aided Design (CAD) software features with automatic name generation. But no matter how a company decides to name its system constituents, the key is to involve all the product development stakeholders. If a naming convention is preferred, it should start from the system level and cascade to the subsystems, modules, etc.

Keeping the Bill of Materials linked tightly to engineering and manufacturing in a two-way relationship is crucial. Any engineering change that impacts manufacturing choices needs to automatically flag the manufacturing engineers who will make the due updates before the release for production. Vice versa, any change request from manufacturing should be carefully reviewed by systems engineers and architects to assess its impact, not just on one subsystem but on the entire Root BOM. They will also check the potential impact of the module interfaces defined by the ICD's. The tight teamwork of all the functions involved in system design is the only way to stay lean in development, cost, and time and maintain a low defect rate.

4.7 MODULAR PRODUCT ARCHITECTURE SUMMARY

Any system's architecture is made of its constituents and their relationships. A product architecture driven by its modules is a Modular Product Architecture or MPA. What sets and MPA apart from other architectures is the simplicity and flexibility of the design, manufacturing, and management of the product. Modules, in the context of the book, are independent assemblies that perform a set of functions. Two keywords in the definition are "independent" and "function." A function is what any entity does. As stated in the chapter, entities, natural or human-made, perform functions to achieve their purpose. When the functions of a system are identified or defined, they can be grouped into sets, avoiding dependencies to have them performed by assemblies or physical entities. Therefore, the module performs a set of independent functions and is, by nature, independent from other modules of the same system. It communicates with them by receiving input and delivering an expected output. Modules can be designed and tested alone in isolation from the system and then be brought in for integration with the rest of the system without compromising the integrity of the whole. Such a property is what makes Modular Product Architecture (MPA) a vital asset for a company involved in the design and delivery of products, processes, services, and even organizations as human systems. Modules, as defined in the chapter and stated above, are functional entities since they perform functions. They are also physical entities, especially for products, because they can be assembled and shipped.

Once a system's modules are defined, they are categorized into instances. An instance of a module is a category or a class of modules. For example, the container module of a bottle is one instance of containers, and a cup as a container represents another instance, as shown in Fig. 4.3. Instances of a module are distinguished from one another through variants. Glass and plastic materials define two variants of the same bottle container instance. Fig. 4.4 shows another example of variants for a city bike. From a generic physical module, the specification of instances and variants is essential for production to respond to specific customer orders.

The definition of a system's subsystems and modules enable their mapping to all the offerings of a company's product portfolio in an overview table defined as a Portfolio Synoptics Matrix (PSM). The PSM helps understand how the functional modules map to the offerings before making any considerations about module instances and variants.

A product portfolio's variants can quickly grow in number, increasing the cost to manage them if not well managed. Methodical planning and design of the modules and their interfaces to be re-used mitigates such an uncontrolled growth. With a clear picture of its products' modules with their instances and variants, a company can make strategic module re-use decisions that will accelerate its new products' releases, increase their reliability while minimizing investments. A new design will leverage known and proven modules. One of the examples shared in the chapter is in Fig. 4.7, where the design of a new impact tool offering took only 1/3 of the time it would have taken in normal circumstances.

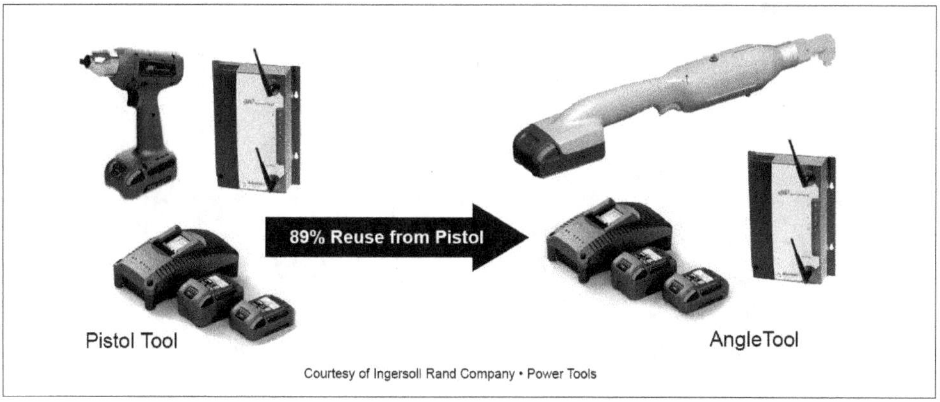

Fig. 4.7 Conversion of a Vertical Tool to a Horizontal Tool

The modules re-use is further enhanced through strategic groupings into Core and Non-Core modules and platforms. The use of platforms is widespread in some industries, the automobile industry in particular.

The chapter offers any company that gets on the *WinnoveX*™ journey through Modular Product Architecture, a better way to organize and protect its critical technologies and investments.

Without an intentional, strategic direction, a company that designs, manufactures, and commercializes various products through specialized divisions or business units will not naturally tend to optimize its cost and product offerings globally. Its organizational structure typically follows divisional level optimization. The resources of a business unit often come from the same industries and are comfortable with their expertise relative to their specific knowledge and experience domain. They do not mostly show a natural inclination to cross domains to globally leverage their expertise by assessing how it applies to other product portfolios in the company as a whole. Fortunately, the creation and adoption of a Modular Product

Architecture provides the appreciation for a broader understanding of all the offerings and new opportunities to accelerate product development, improve reliability while minimizing investments.

5

Systems Design

A company or business is built around a set of core values that directly or indirectly improve people's lives. For a system such as a product, a process, a service, or an organization, companies deliver customer value to satisfy needs that have not been fully met. They are continually working to improve their offerings by identifying new unmet needs or new ways to increase the satisfaction of known needs. Startups are often born around providing solutions for such needs once they are identified and understood. In every industry, competition centers around identifying the potential customers' needs and the definition of better solutions to meet them.

Systems design (when the system is a product) is the process of synthesizing and integrating elements (consisting of hardware, software, and data) to satisfy the identified needs and, in doing so, to deliver value. These principles also apply (Chapter 13) even when the system is a human organization; the needs are defined, goals set, and talents selected to meet them in both the short and long term. In all cases, the organization's purpose or mission drives the right resources' deployment in people and infrastructure to deliver the value.

In general, for systems design, the following steps need to be taken:

- **Step 1**: Identify stakeholders' needs
- **Step 2**: Translate the needs into requirements
- **Step 3**: Select, integrate, and validate the elements whose action will deliver the value expressed by the satisfaction of the needs

These are considered the "Cardinal Steps of Design." Any entity's design always goes through the three steps. Companies that best execute the details resulting from the breakdown of each of the steps win the competitive battle.

Fig. 5.1A shows one example of the main phases of product development, including the cardinal steps described earlier and simplified for the review. Most companies have four to five phases. Some companies establish reviews at the end of each phase to ensure the deliverables are met before approving the investments needed for the following phase.

Fig. 5.1A Product development phases

Fig. 5.1A is the representation that most companies follow. What it does not express, however, is its close link to the stakeholders, especially the buyers, users, and suppliers during the development. It almost appears as if requirements are static, and customers are locked in time until the system is delivered. Companies that have developed a process to engage their customers and integrate their feedback are successful. Chapter 6 covers the case for the battle for innovation. Fig. 5.1B reminds us that buyers and users will flock to acquire systems that show perceived values. Therefore, customers must be considered along the development; if not at least at the beginning and the end of the process. Their engagement will help gauge and identify their unmet needs. Depending on the type of system developed, their feedback should be sought at defined stages of the development. Then, after launch, they will be the happy acquirers of the new product or service.

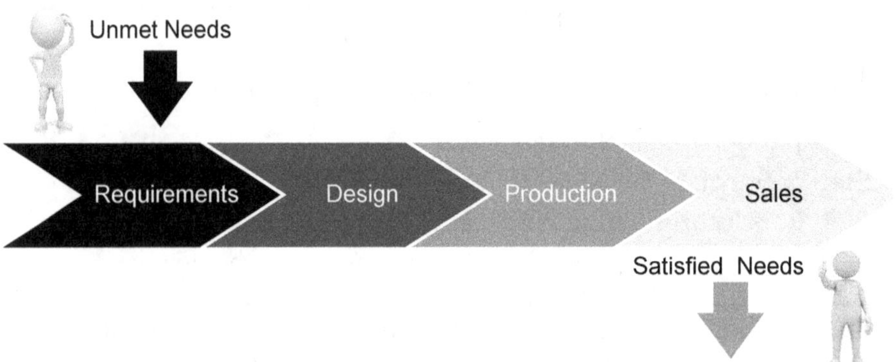

Fig. 5.1B Product Development focused on stakeholders

Considering the customers during the development process that starts and ends with them, as expressed by Fig. 5.1B, Fig. 5.2 better translates my view of the product development flow.

Fig. 5.2 Product Development focused on stakeholders

Fig. 5.2 shows that the stakeholders need to be at the core of every product development. Not only it starts and end with them, but it also involves them during the process. The project team should engage its internal stakeholders at all times.

Since the customers' needs to satisfy are expressed and documented through requirements, it is crucial to dive into both requirements and design phases in the next sections. System requirements definition details are rather technical and are in Appendix.

5.1 REQUIREMENTS

Any successful project delivers its expected outcome(s). In other words, celebrations begin when the designed team meet its requirements and generate the desired results, whether financial or organizational. Success starts with well-defined requirements. The critical question to answer is: "How to define requirements?" Or even harder, "How to define good requirements?" This question comes frequently as project teams are constantly challenged by their leaders to define good requirements.

Requirements are the expression of what the stakeholders need and want from a system.

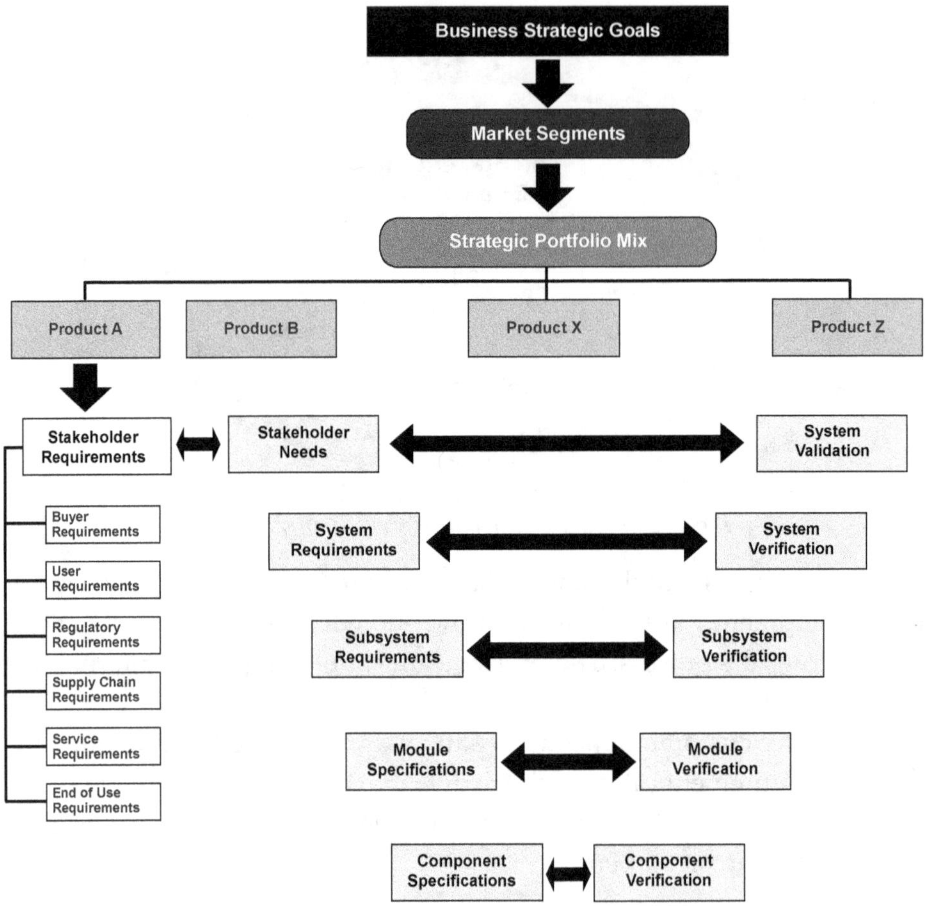

Fig. 5.3 Requirements Flow

Here, stakeholders are all the people who have a vested interest in the outcomes, whether they are customers, users, consultants, engineers, marketers, suppliers, etc. Everyone has a clear interest in the system's success, which could be a device, a machine, a construction, a process, a service, or even an organization.

Requirements being a foundational part of a project, they must be well captured and expressed in clear and unambiguous terms that could be understood by all the stakeholders without further interpretations. Before reviewing the various shapes and forms of requirements, it is essential to place them in the right context.

In general, a company or an organization's strategic goals drive the choices of market segments to address. Then the segments define the portfolios to invest in and which products or services to offer. Only once these critical decisions are made does it make sense to contemplate the stakeholders' requirements based on the identified needs whose satisfaction will be a source of growth. At a high level, the requirements flow is expressed by Fig. 5.3.

Even though this representation focuses on product design, its application covers any system development. Each requirement item of Fig. 5.3 will now be reviewed.

5.1.1 Stakeholders or Market Requirements

As defined above, any person or group of people with a vested interest in the success of an endeavor is a stakeholder. Therefore, the success of any initiative or project is measured by the satisfaction of its stakeholders. Customers will express their satisfaction by flocking to buy the product or subscribe to the service offered. Manufacturing operators are satisfied if the components, modules, and subsystems that make up the system are adapted to the production means with safe and smooth production at a minimum scrap rate. Maintenance technicians will be satisfied if the system is easy to service. Everyone involved has expectations that could be clearly expressed, poorly expressed, or not even expressed at all. The ability of the product managers, systems architects, and engineers to capture such needs and to represent them in terms of requirements will determine the initial success of the product.

It is crucial to capture the voice of the customer (VOC) as the starting point to understand the needs to address, unless the product or service request is unique and belongs to one single customer with precise specifications. We will take a broader view of the customer by extending it to all the stakeholders. The voice of the customer data (assumed to be the voice of the stakeholder data) become an essential ingredient in defining the "Stakeholder Requirements." It is usually assumed that in a company, it is the responsibility of the product management department to collect the voice of the customer, competitive analysis, and other types of market data to define a set of "Must's and Want's" in the market requirements. Product managers should assume the lead in capturing the stakeholders' needs. Still, in general, it is recommended that the requirements definition be a team effort involving as many stakeholders' representatives as possible. The process to capturing VOC is so crucial it became part of some big consulting firm's core competency. During a customer interview, the type of questions asked and how they are asked could determine the outcome of the initiative and lead to its success or failure. There is a real risk in falling into one of two extremes while setting requirements based on a customer survey. Not understanding such a risk can spoil companies' significant investments. Either the interpretation of the survey results leads to requirements that match what is already being offered by competitors or to completely far-fetched and unrealistic requests. In the first case, unless the company is a fast follower with a short development cycle-time, its new product could be obsolete by the time it is released. In the second case, the product may have features for which no one is willing to pay. Therefore, setting the stakeholders' requirements is the first step toward the success or failure of any design project. An executive from a Fortune 500 company, during a product portfolio review, shouted: "Show me the requirements, and I will tell you if we are taking the right road to success!"

The voice of the customer, although particularly important, is only part of the source of the requirements data. From my experience, the customers can mostly provide useful information for incremental improvements. Literature provides references where the VOC is likely to continue to lead along a well-known and existing path.

Most companies do not pursue a complete and full-blown Stakeholders' Requirement Document (SRD), but they mostly create its subset limited to their external customers, known as Market Requirements Document or MRD. Although necessary, the MRD does not cover the internal customers' needs. At the minimum, some fields that should be covered in the SRD as suggested below.

Stakeholder/Market Requirements Key Sections

In this section, the stakeholders' needs, identified through various insights, are internally translated into a requirement document, which is being called stakeholder requirements. In general, stakeholders provide input but not requirements. As always, there are exceptions, but the case considered is for most developments.

Section 1: Business Opportunity

- **The unmet needs**. This section should be clearly stated. It is from the market research summary. Both management and the product development teams should understand the opportunity at hand and the value to the potential customers.
- **The existing solutions on the market**. This section, which is also from the market research summary, should describe the existing solutions on the market and address why the identified needs are still unmet. Such clarification should further highlight why customers are likely to buy the product or service that addresses them.
- **Alignment with the company's strategic goals**. The development of the product or service will require an investment from the company. Therefore, showing how it fits in the overall company strategy and the product portfolio mix is crucial as it helps prioritize the resources available.

Section 2: The Competitive Landscape

- **Megatrends**. Megatrends are likely to shape the market. They will exist whether a solution becomes available or not. They could be technological (some examples are: The Internet of Things, Cybersecurity, Additive Manufacturing, etc.), Political / regulatory (sustainability), Social (Water Scarcity, Poverty), etc.
- **Competitive Differentiation, Positioning and Analysis**. The relative position of the competitors is analyzed based on their Strengths, Weaknesses, Opportunities, and Threats (SWOT). The section should be broad enough to cover the threat of products that come from other applications direct and indirect with the potential to shift the addressable market towards their solutions. For example, cameras on smartphones have changed the photography market forever.

Section 3: Market Segmentation and Data

- **Customer buying behavior**. When customers have a need, any company must understand the product or service's purchase decision process. In the era of e-commerce, a consumer can easily search and be presented with a whole set of solutions, then select the one to buy. Understanding what motivates the selection and final decision from groups of buyers is essential to document. Analytics enable companies to understand the motivation and decision drivers of groups of customers. Learning the phases the customers of a specific system go through, primarily how they evaluate alternative solutions after their landscaping, can make a big difference while conceiving the system.

- **Targeted market segments**. Customers with specialized needs can be grouped in segments. Understanding and documenting such segmentation can lead to better use of resources to target them. It also opens the opportunity to assess how a new system can satisfy other types of needs that the customers may not be currently thinking. The notion of known and unknown unmet needs will be discussed in Chapter 6 as part of *The WinnoveX™ Way* process.

Section 4: The stakeholders' specific needs

- **Buyer requirements**. This section should highlight the product performance requirements. For new electric cars, range in miles or kilometers per charge, time in seconds from zero to 60 miles per hour speed, etc. are some examples.
- **User requirements**. These could be the same as the buyer requirements but not always. The buyer of an elevator or an HVAC unit in a commercial building is often the general contractor responsible for delivering the solution to the building. But the users are the tenants (employees, visitors, customers, etc.). These requirements should highlight the product capabilities expressed as functional requirements.
- **Regulatory requirements**. The regulations and standards whose compliance is needed to sell the product on the targeted market should be captured, and their references expressed unambiguously in this section.
- **Supply Chain and Channel requirements**. Understanding how to acquire, distribute, and supply the product elements and the whole system to the buyers is vital as it could impact its size, packaging, shipping and transportation, and other important features. Channels and distributors' needs are also captured in this section.
- **Service requirements**. The serviceability of a product cannot be just an

afterthought but carefully defined to minimize downtime. Serviceability plays a significant role in the architecture of the product, the interfaces of its modules, size, and weight. In line with the business strategy, the quality of the decisions made regarding how the company services its products will drive future service revenue with a clear definition of the Lowest Replaceable Units or LRU's.[1]

- **End of Use requirements**. Whether a full-blown SRD or an MRD, the document should cover the product's whole life cycle from cradle to end of use. It is possible to design a product with more than 99% recyclability. The design team should pursue the avoidance of components dumped in landfills. Therefore, it should intentionally be part of the requirements.

Section 5: Financial Data

- **Project/Product Development Cost**. This is the section that specifies any requirement about the cost of the project elements or product development costs. Such costs would include resources, prototypes, and pre-production build.
- **Product Cost**. Setting the product cost target helps define the sales price, including all the overhead costs and any margin requirements. Some companies have a detailed cost structure starting with raw materials with all the value-added levels. In such a case, the target cost types need to be defined.
- **Manufacturing and Supply Chain Cost**. The management of the suppliers and logistics, the target of any capital investment, and manufacturing cost will be specified in this section.
- **Marketing and Sales Support Cost**. To be included here are, for example, new channels cost (if applicable), advertisement, sales-related cost (documents like brochures, sales representatives, etc.)
- **Sales and Revenue Projections**. This is an important section. It is always disappointing when investments are made based on some

1 An LRU is the smallest unit (module or component) that may be replaced at designated locations such as repair centers. But when the unit needs to be replaced in the field the term FRU (Field Replaceable Unit) is also used as the smallest unit that can be quickly replaced in the field. For example, a printed circuit board could be an FRU.

sales projections, only to realize that the sales have evaporated after the product launch with the explanation that the market has dried out. The volumes provided by the sales departments are critical, and their providers should be held accountable.

- **Business Case**. A business case can be lengthy and complex, but it can simply be summarized into one statement: if these requirements are met, Quantity Q can be sold with Margin M to increase both the top line and the bottom line by $X and $Y. The business case drives investment commitments. Therefore, at the end of every product development stage, the risks to its attainment are assessed with their mitigation plans to adjust the outcomes. A decision is made to pursue, put on hold, or cancel the project.

Section 6: Schedule

- **Project/Product Development Schedule**. The significant milestones necessary to release the system per company's strategic goals define this section. The data in this section is particularly important if Earned-Value[2] management is in place in the company.
- **Launch Schedule**. The company sets every product's launch date to drive the project to closure. If multiple releases are planned, each release date should be communicated and documented.

5.1.2 System Requirements

Once the market or stakeholder's requirements document is released, the system engineers and architects have the duty to translate it into a technical or design language to create the system requirements document or SyRD. While the colleagues from product management or marketing may identify essential functions that need to be performed by the new product, the specification of how well they need to perform should come out of an agreement between marketing, engineering, manufacturing, and supply chain. Some delicate iterations are necessary to converge the different stakeholders toward the metrics that represent the market requirements' intent to define realistic expectations. It is to be noted that each requirement comes at a cost. There-

2 Earned Value tracks actual work vs. project plan and work completed value. Based on the completed work, Earned Value shows how much of the budget and time should have been spent.

fore, unnecessary requirements thrown in the document by marketers or engineers can compromise the cost competitiveness of the system and delay its release. Generally, once the requirements are established, they are frozen, and the only corrections allowed should be due to a significant market shift. System requirements, as stated above, are the translation of the stakeholders' requirements in the language understood by those who will develop the system. It also integrates the inputs from prior analyses such as Design and Process Failure Modes and Effects Analyses (DFMEA/PFMEA) as well as other functional analyses. System requirements with their verification and validation (V&V, meaning verification of the requirements and validation of the needs) need to be defined simultaneously.

Requirements definition is the most critical phase of any system design. If missed, a system may be designed and released, but it will not generate the financial return on investment.

The existing literature can be somewhat too generic in defining the system requirement documents' content. The documents include the functional and non-functional requirements (as doing and being of the system) and can also be categorized as performance and operational requirements followed by design constraints. Requirements often presented under different types and formats can make them complicated to grasp fully. They are complex but should not be complicated. Fig. 5.4 gives a view that shows the kind of requirements and their relationships.

	Functional Requirements	Non-Functional Requirements
	What the system does	What / How the system is
System Functions	√	–
Performance Requirements	√	√
Operational Requirements	√	√
System Features	√	√
System Constraints	–	√
Other Requirements	–	√

Fig. 5.4 Technical Requirements Types and Relationships

The Stakeholder Requirements Document (SRD) and the Market Requirements Document (MRD) put any development project in the business context. Their definition should then drive the creation of the System Requirements Document which is a design document.

ΔP A full section on system requirements content and allocation to subsystems and modules is in Appendix C.

5.2 INTERFACE CONTROL DOCUMENTS

NASA defines an interface as "A design feature of a piece of equipment that affects the design feature of another piece of equipment."[3] Even though modules are independent entities, they communicate with one another through their interfaces. NASA's document continued to define the purpose of an interface control: "It is to define interface requirements to ensure compatibility between interrelated pieces of equipment and to provide an authoritative means of controlling the design interfaces."[4]

As presented in Chapter 3 (Section 3.5), Interface Control Documents' (ICD) definition occurs the first time as requirements are being defined, and the design teams agree on how their different modules will work with one another. Such a collaboration prevents unilateral changes that can compromise the whole system.

Interface elements may be mechanical, electrical, or informational for software. When they are:

- Mechanical they could be about physical shape, dimensions, location (form and fit), forces, etc. Mechanical interfaces are better represented through drawings and documents with detailed comments.
- Electrical, we could consider impedance, circuit logic, signal characteristics, electromagnetic influence, data transmission, etc.
- Software, they could be communication protocol, message structure, and flow, error management, etc.

---

3 NASA Reference Publication 1370 of January 1997

4 Ibid.

5.2.1 Change Control Board

Interfaces are so critical to the interoperability of modules that any lack of change control would lead to a disaster. Any investment into standard interfaces can be spoiled without a disciplined change control governance (process, board, cadence, etc.). Once created, ICD's need to go through an approval process that involves all the stakeholders. After a thorough Independent Design Review (IDR), the document should go under configuration control. A Change Control Board (CCB) should be established to maintain its integrity and assess its impact on all interfacing modules before any new change is released. Any Change Request (CR) should be directed to the CCB leader. They should be collected and updated with particular attention to forward and backward compatibility and change impact on overall system integrity.

ICD's are substantial parts of any system requirements. Thus, any design review should include compliance with all interface requirements.

5.2.2 The Future of Requirement Definition in the 21st Century

The 21st Century is shaping up as the era of a full-blown digital world with smart and connected systems, big data, artificial intelligence, etc. In such a world, existing machines provide themselves a clear picture of their usage and health status. The data define a specific customer usage profile. They can lead to the grouping of customers with similar usage profiles with products or services designed for their particular needs (meaning their requirements). This new capability will shape how to define requirements in the future, and modular design will play a significant role in industries to contain the cost while getting the system remarkably close to each customer. It is de facto the era of mass customization. Companies that develop the capability to meet individual customers' needs with their products, without the proliferation of design requirements but from an established set, will win on the marketplace. They will offer customized products at a competitive price at a lower cost without compromising their margins. In the digital era, heavy use of data analytics and modeling and simulation will change how requirements are set and how products are conceived and delivered, including design, manufacturing, logistics, installation, service, and ultimately recycle at the end of use.

5.3 SYSTEMS DESIGN PROCESS

Since each company has its own systems design process, a deliberate choice has been made during the activities of this book not to dive into the details of the process. Whether a waterfall or an agile process or a combination of both are adopted to design products or, in general, systems, a few observations are essential for companies to note:

A. Product development phases are not linear

Referring to the example of product development phases summarized in Fig. 5.1A (re-inserted below for reference)

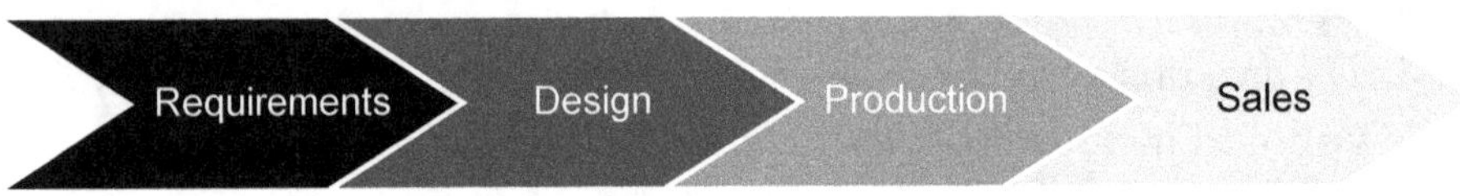

Fig. 5.1A Product development phases

Or Fig. 5.1B that shows how the process starts and finishes with the customers,

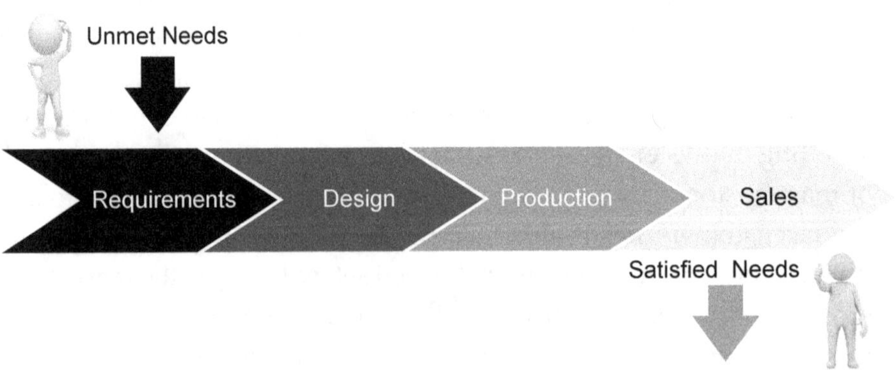

Fig. 5.1B Product Development focused on stakeholders

I said then, that a better representation should be by the loop in Fig. 5.2.

Fig. 5.2 Product Development focused on stakeholders

Constant reference should be made to the stakeholders (external and internal customers) to keep them engaged in system development.

B. The market environment is not static but dynamic

Considering the system development phases in a loop rather than linear is just the beginning. Another important observation is that the market environment is not static but dynamic. Since most developments could take months or even years, by the time the system being designed is released, the market may have evolved. Other systems may have been released, bringing more knowledge to the customers who may have new expectations. Also, new technologies may have been introduced with better performance than the ones considered. The product development phases shown in Fig. 5.10, represent such market behavior.

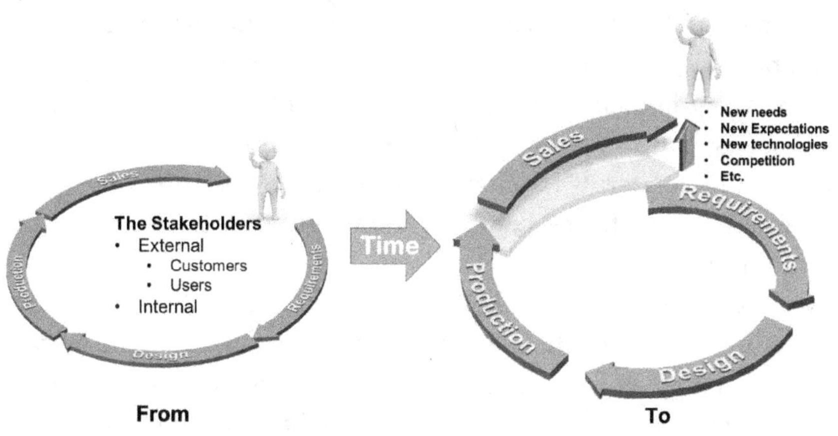

Fig. 5.10 Product development phases with the dynamic nature of the market

The product development environment is in constant and rapid evolution. Some design projects with long development cycles face the risk of launching products that are obsolete before they are released. When the companies operate under lower speeds, a modular approach to a product design is critical to market success. It cuts the development time by reducing complexity and, more importantly, by driving innovation at the module level rather than at the full system level. Having company or industry-standard interfaces allows the rapid insertion of innovative solutions on a short cycle. The adoption of industry-standard interfaces or the partnership with other industry players to jointly develop interfaces opens new opportunities to speed development time. Companies can either leverage well tested and validated industry strategies and methods to connect modules or share risk and cost with other partners to define common interfaces. Communication protocols are a typical example of widely shared interface elements. Time is an important factor to be considered when developing new products.

Frequent releases keep customers engaged and competitors struggling, especially if they need to make significant changes to their products every time to compete.

5.4 SYSTEMS DESIGN SUMMARY

To stay on the leading edge or to be relevant and competitive, companies need to create and sustain the culture to invest in the design of new offerings. The success of any system design starts from having good requirements. Not taking the time to define them will sow the seeds of a future failure.

This chapter emphasizes what makes good requirements. It covers the main fields of a requirements document related to both internal and external customers' needs. It also highlights how the needs should be characterized and translated into the terms familiar to the design community to execute.

The information flow from a company's strategic goals to various requirements is summarized in Fig. 5.3.

A few words were also shared about requirements in the 21st Century, characterized by a full-blown digital world with smart and connected systems, big data, artificial intelligence, etc. It offers opportunities for a new approach to systems design. Data analytics are now revealing golden customer and product information that company systems have held for a long time. Existing machines can now provide a clear picture of their usage and their health status in general. Such data can help define specific customer usage profiles and lead to their grouping and a better definition of their segments and needs. Customers with similar usage profiles can have products or services designed to address their specific needs. This new capability will shape how to define requirements in the future.

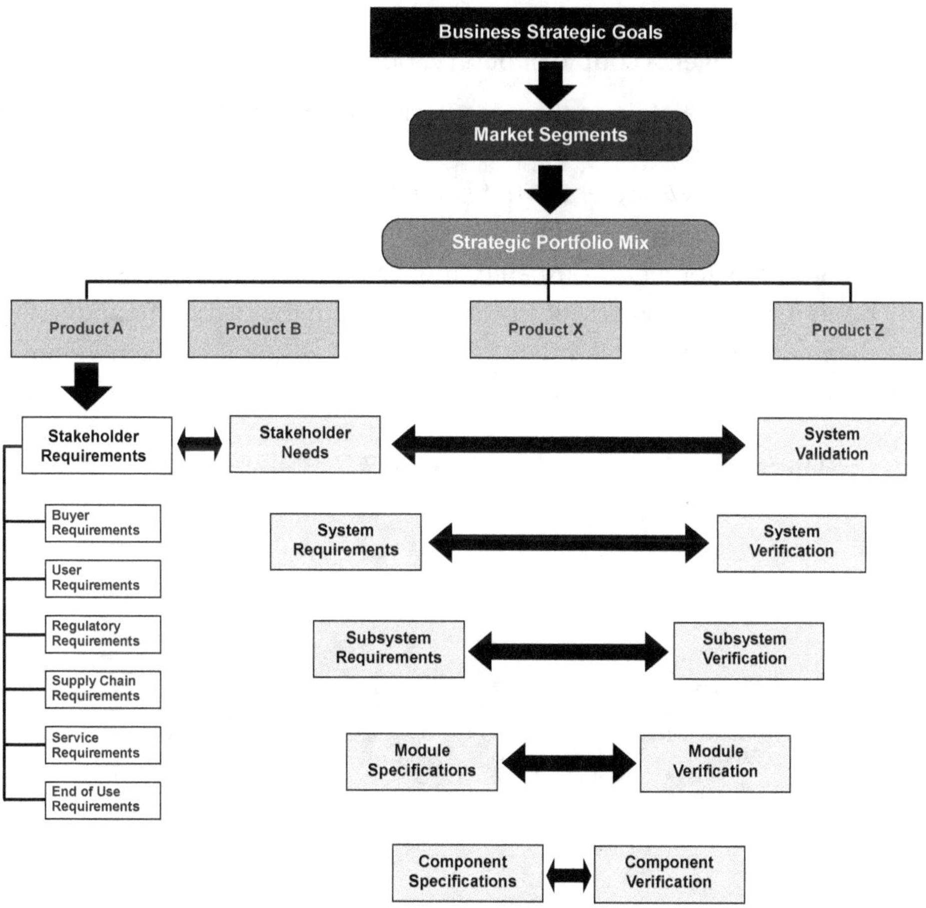

Fig. 5.3 Requirements Flow

The modular design will further augment it and facilitate new opportunities. It will play a significant role in companies to contain cost while getting their products remarkably close to each customer's need, opening the door to mass customization.

The companies that develop and master the modular approach to meet individual customers' needs with their products from an established set of modules and platforms create sustainable differentiation in the marketplace. They win by offering customized products at a competitive price and at a lower cost without compromising their margins.

Regarding the systems design process, a deliberate choice has been made not to dive into the details of any process since they are specific to each company. However, a couple of observations were worth noting.

A. The product development process is not linear as it should start and end with the customer.
B. The market environment is not static but dynamic as new products or technologies continue to be released by competitors, setting new expectations in customers. Companies must either shorten their development time to market or keep getting feedback from the market to understand the relevance of the requirements they are designing to meet. If neither is done, they face the risk of giving birth to obsolete products or systems.

Both observations are synthesized by Fig. 5.10 shown again below.

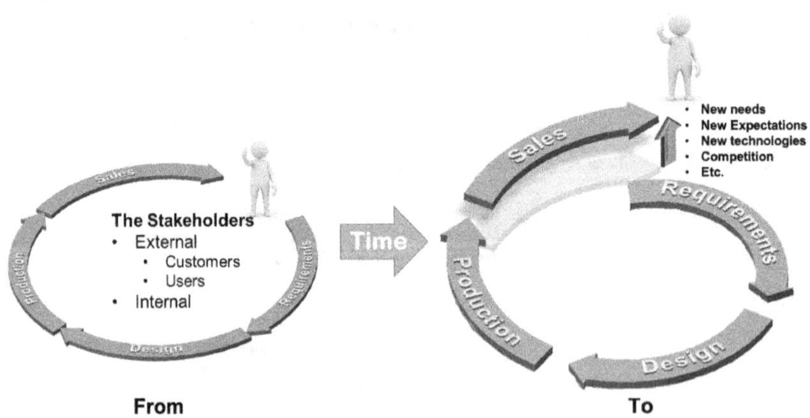

Fig. 5.10 Product development phases with the dynamic nature of the market

Modular Product Architecture provides the opportunity:

- To cut development time by reducing complexity and, more importantly, by driving innovation at the module level rather than the full system.

- To leverage company or industry-standard interfaces to get a rapid insertion of innovative solutions on a short cycle. Frequent releases keep customers engaged and competitors struggling, especially if they need to make significant changes to their products to compete.
- To have systems' requirements clearly allocated to the subsystems, the modules, and even to the components, enabling their design and manufacturing anywhere without compromising the integrity or performance of the products.

6

The WinnoveX™ Way

Several dictionaries define innovation as the act of introducing something new. It could be a new product, a new process, or a new service introduced to solve a known problem or an unmet need. But in the 21st century, the idea that innovation brings solutions to a known problem needs to be amended. The innovators today need to understand that innovation brings solutions to conscious and unconscious needs. The conscious part is what we would call "known unmet needs." But since not all needs are known, we would call the unconscious needs "unknown unmet needs." In this scenario, people do not understand that those needs existed until they see the solution. That is the new way of seeing innovation. Human needs are complex and should not be explored the same way. To manage its resources and strategies towards sustained innovation leadership, a company needs to understand the distinction between these two types of human needs. Two key questions to ask are:

- How does a company uncover unknown unmet needs?
- How can a company consistently tackle both known and unknown unmet needs to bring novel solutions to the marketplace?

Characterizing the unknown unmet needs as "induced needs" generated countless discussions with people arguing that a need cannot be induced. Until the electronic book came into the marketplace, no reader would have expressed the need to have a device that would contain numerous books or a whole library. For centuries people were happy to read their books in hard copies. Readers could carry them, mark them, or highlight the passages they wanted to review at a later time. The need for reading was fully satisfied with the existing solutions. Today, however, millions of people carry mobile digital libraries in various forms, with an ever-expanding worldwide market.

Car heated seats are another example of unknown unmet needs. Cars existed for decades before Robert L. Ballard, a General Motors employee, filed his car seat heater patent in April 1951. If GM took a survey about the concept of heated car seats to determine the consumer's interest, people, mostly from harsh winter regions, would have recognized the innovation's value. But this unmet need lay dormant for years before they installed them on a Cadillac in 1966.

Based on several innovations like the two examples presented, unknown unmet needs can be further classified into two categories: Comprehensible unknown unmet needs and incomprehensible unknown unmet needs.

- Comprehensible unknown unmet needs – these needs are not evident to people. But they can be understood, when expressed, because they are straightforward or related to a reference. A heated car seat is an example of such a need. If asked about having a heated car seat, the need, although unknown would have been comprehensible.
- Incomprehensible unknown unmet needs – these needs are not understood at first because there are no references to help the person understand them. The need for such systems is difficult to discover out of customer surveys. The systematic way to identify these needs is mostly through careful observations and studies of human struggles to overcome some natural hurdles supported by functional analyses. When not using functional analysis, the discovery is most likely accidental.

When incomprehensible unknown unmet needs are expressed, they are typically met with skepticism. Their understanding by potential customers takes time; therefore, their solutions must be handled with the necessary care they require to be brought to market. Companies that pursue them must be aware of all the hurdles to overcome and risks to mitigate for their commercialization.

Let us review an example. One of the most interesting facts about our human journey through history was the Germ Theory of Disease. The theory stated that certain diseases were caused by microorganisms known as germs that were too small to be seen with bare eyes but could invade the human body. They could only be seen through a microscope. Until Louis Pasteur, the French microbiologist and chemist confirmed the theory in mid 19th century and invented the pasteurization process to treat milk to stop bacterial contamination, the need to do so was unmet, unknown, and incomprehensible.

The second question raised at the beginning of this chapter was how companies could consistently tackle known or unknown unmet needs and address them with novel solutions.

There are a lot of methods, tools, and processes in use that address, one way or another, the category of known unmet needs. Through rapid development cycles or modeling and simulation, companies have many ways today to validate known unmet needs. However, it is not the case for unknown unmet needs, so we will focus on them to help companies seize the opportunities they bring, including a substantial competitive advantage when they are well executed.

As if the various categories of innovation types presented in this section of the book were not complicated enough already, another dimension shall be added as companies innovate to address both known and unknown unmet needs. It is the dimension of incremental or radical innovative solutions. The complexity of known and unknown unmet needs, with the dimension of incremental and radical solutions, can be simplified in the two by two matrix shown below in Fig.6.1.

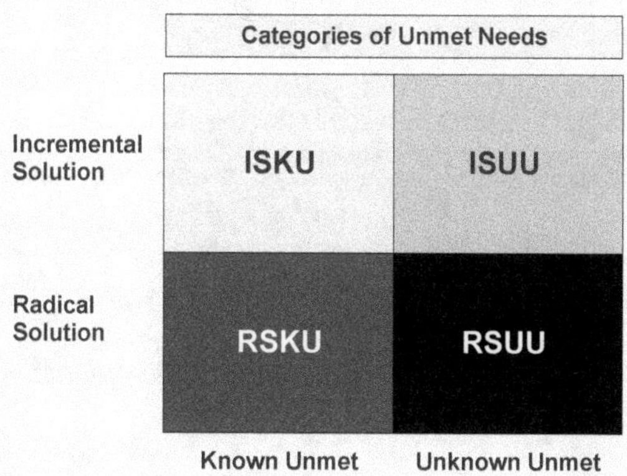

Fig. 6.1 Unmet needs with solution approaches

Six categories emerge from the previous paragraphs' considerations for the designers or architects and all those involved in systems innovation. We will assume that a known need should be comprehensible when expressed to stakeholders. If it is not understood, its description should be formulated again. The formulation of needs is particularly important since it leads to specific requirements and a design that can validate them. An inadequate formulation of requirements is a recipe for disaster and a waste of precious resources, as discussed in Chapter 5.

If an unmet need is unknown to the stakeholders when expressed, they can understand it if there was a prior reference point. In the absence of anything to relate to, the need may not be comprehensible. Once a defined need is categorized as comprehensible or incomprehensible to the potential customers, a company may decide to pursue an incremental or radical solution. In the next sections, we will briefly review each of the six categories identified based on the characteristics of unmet needs and the solution approach a company may adopt.

Fig. 6.2 summarizes the structure of the unmet needs to consider.

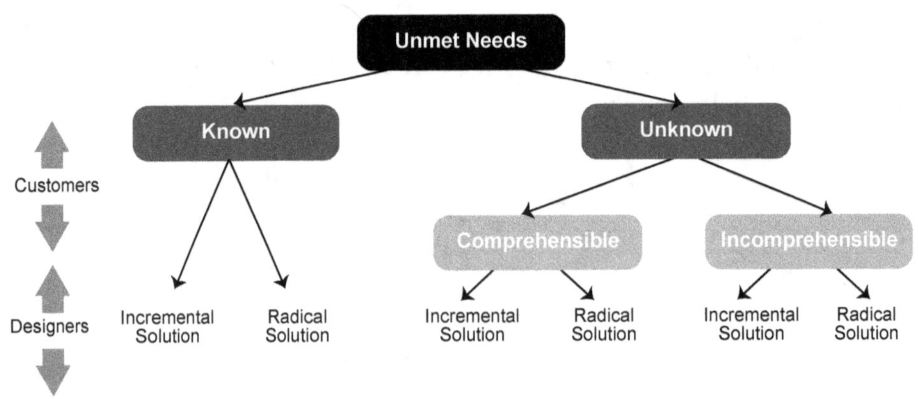

Fig. 6.2 Unmet needs categories and solution types

The observation of the cases presented in Fig. 6.2 leads to a few lessons.

6.1 KNOWN UNMET NEEDS

Companies need to understand which needs they focused on investing. Whether the solutions for known unmet needs are incremental or radical, they fill a felt, somehow desired void. People know what they need, even if there is no solution or if the existing solutions are unsatisfactory. Let us take a closer look at incremental and radical solutions.

6.1.1 Known Unmet Needs with Incremental Solutions

Incremental solutions to known unmet needs (ISKU) are usually the result of customer surveys. Innovative companies should have the standard practice to set up transactional feedback and formal surveys of their customers, users, and any other stakeholders to continuously improve the systems they deliver. When people are asked what they would improve on a product, a process, a service delivery, or even an organization, they express their pain points or desire for improvement. Issues presented that are likely to benefit a whole market segment should be incorporated into existing systems incremental improvements. A market is a dynamic place. Without a minimum engagement to get from those interacting with a system, inputs on how it could be improved would miss the fundamental principle that drives customer loyalty: showing that their voice is heard.

ISKU (incremental solution to known unmet needs) needs are critical to the competitiveness of any company. They drive the constant flow of steady improvements that customers appreciate as the response to their immediate desires or expressed needs. Even though they may not give a considerable edge over a company's competitors, they keep loyal customers from going somewhere else. Their impact is not negligible as they guarantee a steady revenue flow until the market takes a new direction. The revival of hybrid cars as a solution to air pollution and gasoline price fluctuations due to oil production crises is an incremental solution to known unmet needs. The concept of cars running on two different sources of power is not new. Even though a gas/electric car was introduced at the turn of the 19th century by Ferdinand Porsche, an Austrian engineer, it did not gain traction due to Henry Ford's competitive gasoline-powered cars. But the growing air pollution in large cities due to massive adoption of vehicles as the preferred form of transportation, and the oil crisis of the early 1970s, pushed companies like

Toyota to heavily invest in new technologies and introduce the hybrid Prius in 1997. Many other hybrid offerings followed in the auto industry. We do not have to go far to find examples of incremental solutions to known unmet needs. They are all around us in our homes, from quiet appliances, cars with powered windows, to self-propelled lawnmowers. These innovations are the results of tireless development teams that guarantee a continuous improvement of their products or processes.

6.1.2 Known Unmet Needs with Radical Solutions

A known unmet need can also be satisfied with a radical solution. It is the RSKU need (radical solution to a known unmet need) from Fig. 6.2. In this case, even though the need is known to the stakeholders, they may not make concrete suggestions that could lead to a radical solution. A customer survey is still necessary since the need is known, but it is relatively rare for customers to suggest a radical solution. Even if they are bumping into the need every day and feel or often express it, the radical, innovative solution may still catch both customers and competitors by surprise, thereby putting the system on a new path or a direction different from what people expect. The innovators gain a clear competitive and time advantage in the market while others struggle to catch up. Henry Ford's affordable car Model T example, mentioned in the introduction, radically solved the known unmet need to clean the early 20th century's big cities from severe pollution from horse manure. Microwave ovens are another example of a radical solution to a known unmet need. The need to accelerate food cooking was always present. The introduction of a countertop appliance that reheats foods in seconds with cooking times estimated in minutes was very appreciated and is now in millions of households worldwide.

A lot is written about various customer survey techniques in the existing literature, and there are excellent companies specialized in the field. However, to get to a radical solution to a known need, the survey questions must be structured to have answers that can later be connected to an excellent functional analysis. The process will be reviewed in Chapter 11 when presenting Product Management.

6.2 UNKNOWN UNMET NEEDS

When unmet needs are known, it is straightforward to consider whether to address them with incremental or radical solutions, as shown in the previous section. In the case of unknown, unmet needs, the design team must assess whether the unmet need is comprehensible or incomprehensible to the potential customers before determining to use the incremental or radical approach. Four categories of needs emerge from the considerations made above:

- Unknown, unmet needs comprehensible to the customers and addressed with incremental solutions,
- Unknown, unmet needs comprehensible to the customers and addressed with radical solutions,
- Unknown, unmet needs incomprehensible to the customers and addressed with incremental solutions and
- Unknown, unmet needs incomprehensible to the customers and addressed with radical solutions
- Such distinctions are necessary for better management of a company's investments.

6.2.1 Unknown Unmet Needs (Comprehensible) with Incremental Solutions

This section covers the category of an unknown, unmet need but comprehensible to the customers. The company can then choose to address it with a solution deemed incremental by the customers. This case was mentioned earlier as an ISUU need (Incremental Solution to an Unknown Unmet need). The only twist added is that the need is comprehensible to the customers when expressed. The formulation of this type of need is important. It must be well described in terms that the customers can understand and embrace. After passing such a hurdle, the incremental solution can be like any known unmet need. A customer survey would also be vital to unlocking the potential for continuous improvement.

Over several years, I observed that the simple notion of an incremental solution to an unknown, unmet need comprehensible to the customers baffles people. The question arises how a solution to an unknown, unmet need can be addressed through incremental innovation. It is about needs that have not yet been met and which the customers ignore. An example would help grasp the notion.

In 1996, Kone, a Finnish elevator company, released a new elevator design that did not require a machine room on top of the shaft.[1] Until then, electric traction elevators needed a room on top of buildings to accommodate their hoisting machinery and electrical equipment.

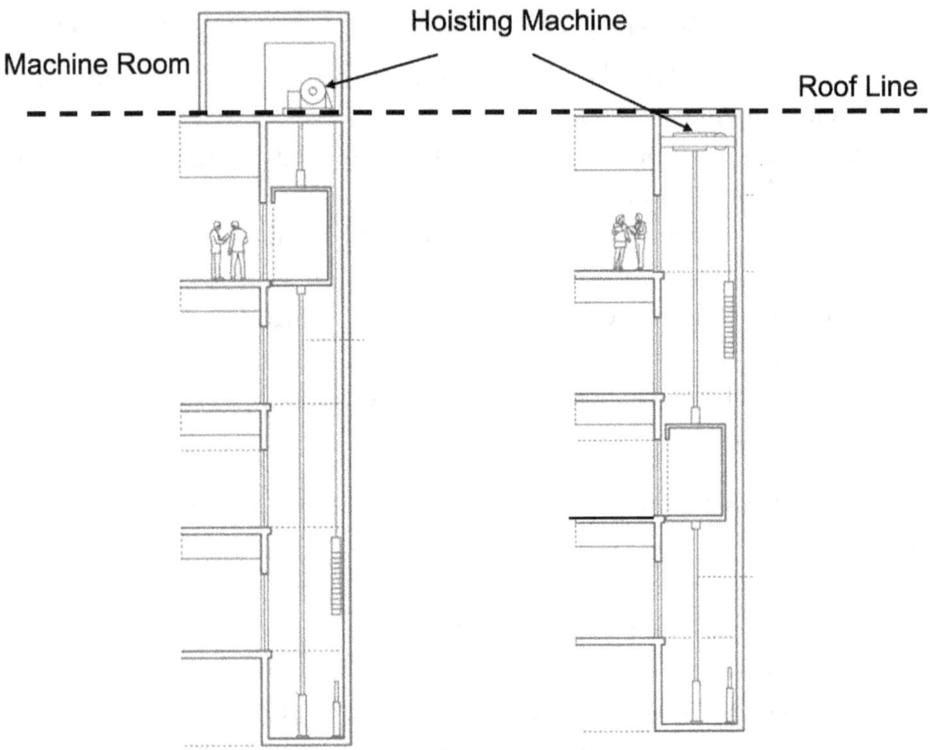

Fig. 6.3 Illustration of elevators with and without a machine room on top of their shafts

When Kone launched its new elevator design under "Monospace" commercial name, the building architects, who never complained about creating a machine room, loved the new solution as it simplified the design of the top of their buildings. Once the solution was unveiled, the need that may have always existed but was unknown or unconscious became apparent. As can be seen from illustrative Fig. 6.3, there is nothing above the roofline.

1 The shaft is the vertical space reserved inside buildings to allow elevator cabins to go from floor to floor.

Kone's solution collocated the hoisting machine inside the shaft keeping the roof free. The solution gave the building architects the freedom to play with their roof design without any other construction constraint on top of the building. Even though the need was unknown, the solution was still incremental to the overall elevator function as a building transportation vehicle. The competitors soon followed by creating their versions of the solution with various designs and technologies.

6.2.2 Unknown Unmet Needs (Comprehensible) with Radical Solutions

This category was discussed earlier as an RSUU (Radical Solution to an Unknown Unmet) need. Like the previous category, even though it is assumed that this type of need is comprehensible when explained to customers, the market acceptance of their radical solutions may not be straightforward. Needs that customers never imagined before, addressed with radical solutions, require the users' education since they may have to acquire new skills or appreciate the problems solved. When in 2001, Dean Kamen introduced the Personal Transporter (PT) manufactured by Segway company, nobody imagined the need for a perfectly self-balanced two-wheeler for campus or city travel. There was a big hype before its release. Several people estimated its impact as massive as the introduction of personal computers. Others predicted it would be as big as the Internet itself. The company estimated its sales to reach hundreds of thousands per year, but the sales predictions proved dismal. Only around ten thousand sold in a decade. The need to have a new personal transporter different from a bicycle or a skateboard was not evident to people, but they understood it after seeing it. The radical solution was stable and could move fast. After overcoming my initial hesitation, I tried the product, and it was incredibly useful. I could move with ease. So, why is it that it did not meet the company's sales projections?

There is a distinction between getting the customers to understand an unknown, unmet need when presented to them and getting from them enough suggestions to derive a radical solution. Since they do not even know they have such a need, comparison to existing devices is often made with the risk of oversimplifying the expression of the need. Once they understand it, their attempt to offer a solution would naturally go toward incremental solutions. So, the additional difficulty here is that a customer survey would not be enough to lead to a radical solution since the need was unknown from the start. The analysis of the functions performed by the existing solutions coupled with a good landscaping of all the potential applicable technologies is the systematic way to get to radical solutions. Therefore, if the investigating company wants to break away from its competitors' pack, it is recommended to add observation sessions to the formal surveys. The observations should be conducted in sessions where the customers know they are being observed as they deal with the system and in sessions unbeknown to them. Trained eyes watching customers or users in action with the product, can yield surprising discoveries and help identify the functions being performed.

6.2.3 Unknown Unmet Needs (Incomprehensible) with Incremental Solutions

We are now in the category of an unmet need incomprehensible to potential customers. Addressing such a need with an incremental solution may present some unforeseen challenges. If the solution is not differentiated enough from other existing offerings the innovating company would take upon itself to pursue a need with the risk of a market that may not appreciate the solution's full value. With the proposed solution being incremental, it is highly recommended that companies be very cautious about pursuing innovations in this category. They must know that they are in for a long game that could consume precious resources of all kinds with a return that may not be worth the investment. There would be an investment for the solution and another investment for the customers' education for its adoption. In conclusion, an incremental solution to an unknown, unmet need not readily comprehensible to the customers may not be worth pursuing. We could argue, but Nintendo's Wii Vitality Sensor was very intriguing to me, a buyer of the company's products, when it was announced in 2009. A sensor that would inform the players about their body and relax them during their game time? Was that really a need? It was not clear why the company would pursue such a solution. The biggest unknown that made it intriguing was its integration into the other video games, which the company did not comment on. It was an incremental solution to an unknown, unmet need that some of my friends and I could not fully grasp. Eventually, the company did the right thing and did not commercialize the product. No one was surprised when they announced in 2013 that it was canceled.

6.2.4 Unknown Unmet Needs (Incomprehensible) with Radical Solutions

This category covers unknown, unmet needs incomprehensible to the customers, which a company decides to address with radical solutions. These types of unmet needs present two challenges. First, they are incomprehensible to potential customers when expressed to them because they have no reference point. Second, the solution approach is radical, meaning that it is not incremental to an existing offering. Generating radical solutions in this category is a higher investment risk than the other types. At the same time, it presents a potentially higher reward. When this type of unmet need was mentioned in this chapter's introduction, the Germ Theory of Disease example was given. The notion that a disease was caused by tiny microorganisms known as germs that invade the human body and could only be seen in a microscope did not make any sense. Only when Louis Pasteur demonstrated, through his invention of pasteurization, that milk fermentation could be processed to eliminate any bacteria, the theory was understood. Before Pasteur, efforts deployed three centuries earlier by Girolamo Fracastoro, an Italian physician who lived between 1478 and 1553 and later in the 19th century by Friedrich Henle, a German pathologist (1809-1885) were met with skepticism. The time's prevalent beliefs were too strong for people to see anything else, even though the need to treat milk from fermentation to prevent diseases was real. Such an unmet need can be uncovered through careful behavioral observations and functional analyses, which require time and resources.

6.3 COPING WITH UNMET NEEDS

The six categories we just reviewed have been summarized in Fig. 6.4 to allow *The WinnoveX™ Way* adopters to assess which need class they are facing before addressing it. Such an assessment is necessary to establish the resources required to get meaningful results. The question becomes what the company's focus category should be. The short answer is all of them.

The threat of a company being disrupted by various competitors comes from different directions. Conventional competitors come from the same domain and are focused on the categories of known unmet needs based on their customers' feedback or the category of the unknown, unmet needs but comprehensible to customers. Meanwhile, companies whose system offerings serve other markets could all of a sudden realize that their technologies have the potential to bring some unforeseen benefits to customers of another business domain. When Sony Walkman, a portable cassette player, was released in 1979, it was a "must-have." Anyone could listen to music on the go. During its product life span lasting until 2010, the company sold nearly 200 million units worldwide. Several companies released their own product versions. Sony went on to release its portable digital player based on compact discs or CDs in 1984.

The new product was an Incremental Solution to a Known Unmet (ISKU) need. Everything seemed to be set for Sony to continue its dominance of the portable music players market, but in 1998 the MPMan was launched from South Korea by Saehan Information Systems. It was the first portable digital audio player based on mp3 files. Other companies followed with their version of MP3 players. In October 2001, the computer maker Apple Inc. released the iPod. Apple made the product beautiful and simpler to use from a hardware perspective and then added the iTunes store to sell the music --- so it was a radical solution by integrating hardware with software seamlessly. The iPod took off, and the company has introduced what was a Radical Solution to a Known Unmet (RSKU) need. Radical innovations tend to introduce a new generation of solutions and must be carefully thought through without rushing to release them. In fact, Steve Jobs' product came out three years after the MPMan. Without a thoughtful and good execution before release, radical solutions have helped fast follower companies that did not make the initial discoveries. Still, they have benefited from the ill implementation of their first to market predecessors.

I like to tell people that innovation is like reaching out into the future to bring things to the present or merely accelerating the present into the future. An invention is successful when it becomes so much of people's regular lives to wonder how their life was ever possible without it. New habits are created because of humans' adaptation to innovation. Ultimately a new solution is successful when it substantially impacts people's lives, driving a broad adoption and a significant Return on Investment (ROI). However, companies do not always use the right methods to drive innovation to pursue breakthroughs, thus lowering their return on their investment.

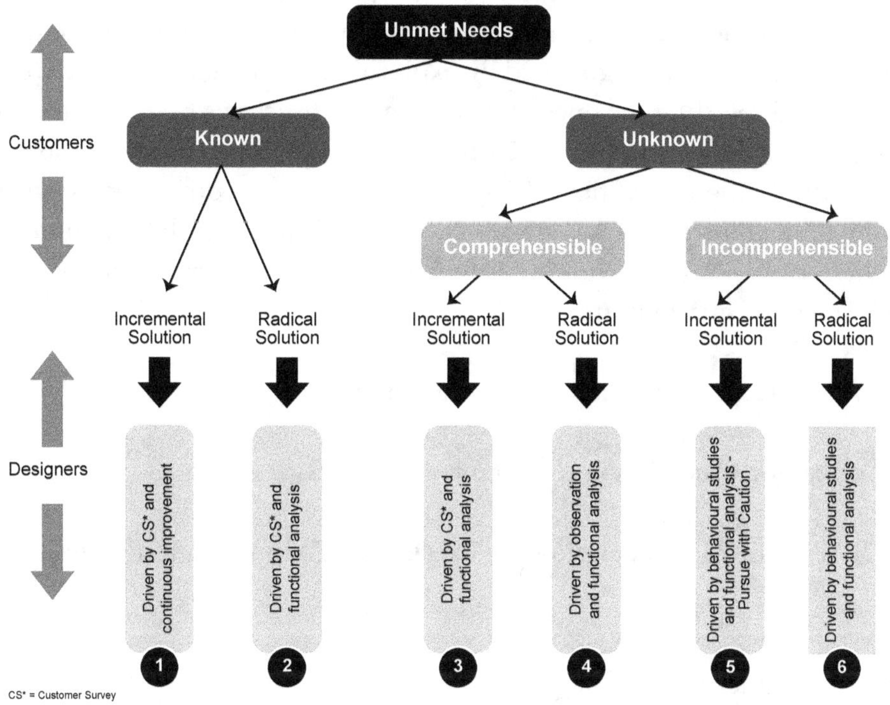

Fig. 6.4 Unmet needs categories and solution types with their drivers

Fig. 6.4 extends 6.3 further by adding some discovery approaches.

6.4 THE WAY TO WINNING INNOVATIONS

This section of the chapter describes two notions. The first notion is a common denominator to all the categories of innovation illustrated in Fig. 6.3. The second is to present *The WinnoveX™ Way* process.

Throughout the book, it was described that every entity, whether natural or human-made, performs a function or set of functions. We've extensively covered functions through the previous chapters. I have explained how they are identified or defined for existing or new engineered systems. They are the foundation for modules creation, and as it was presented, modules are the building blocks of modular systems.

Understanding the functions an existing entity performs or a new one will perform, in other words, understanding what a system does or will do, is key to innovation. Once the functions and modules of a system are defined, many solutions can be engineered by exploring a landscape of various technologies and selecting the ones that perform the functions best. The following equations are defined as the **Fundamental Equations of Innovation.™** They apply to the innovation of any system (products, processes, services, and organizations as human systems). The application of these equations also defines the Way to Winning Innovation Excellence or *The WinnoveX™ Way*. The two equations are:

(1) Functions + Technology = Innovation

And taking execution into account, it becomes:

(2) (Functions + Technology) x Execution = Winning Innovation

Assuming:

- **F = Function**
- **T = Technology**
- **E = Execution and**
- **WI = Winning Innovation,**

Equation 2 can be formulated as:

(2B) (F+T) x E = WI

The second equation should be read as: Good execution of technologies that perform value-driving functions leads to winning innovations. *The WinnoveX™ Way*, it defines, can be represented by Fig. 6.5.

As simple as they may seem, equation (1) addresses innovation while equation (2) describes winning innovations. They amplify each other to lead the way towards *WinnoveX.™* A critical requisite to gain the full benefit of equation (2) is to deploy a modular product architecture based on excellent functional analyses. The requisite will become evident in the next sections and chapters.

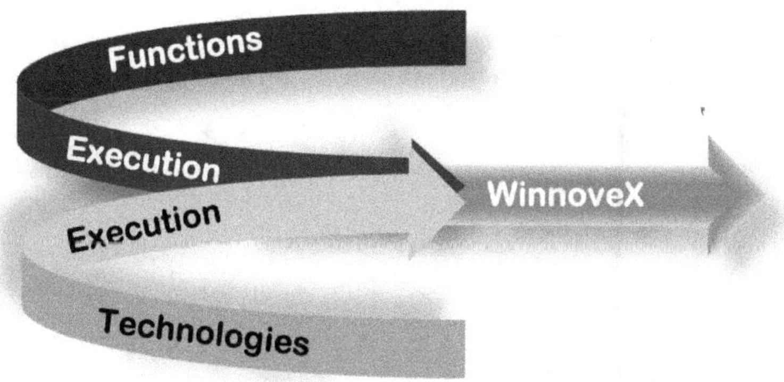

Fig. 6.5 The WinnoveX™ Way

Equation (2) has raised a couple of important questions every time it was shared:

- Why do you need to add execution to the innovation equation to make it the winning innovation equation?
- Why do you consider execution to be so essential to make it a vital element of innovation success?

Even though functions are at the core of *The WinnoveX™ Way*, every system needs to meet functional and non-functional requirements. Execution addresses the non-functional requirements such as aesthetics, cost, etc. It is crucial to delve deeper into execution to assess its criticality to achieving winning innovation excellence.

Although every effort was made to avoid mathematics in the book, the graph below is used to express the value and importance of execution by plotting winning innovation as a technology function.

In Fig. 6.6, equation **(2B) WI = FE + TE** has been graphically represented with the meaning of the letters expressed earlier:

- **F = Function**. Functions evolve very slowly over time. Therefore, for the sake of the exercise, they would be considered constant.

- **T = Technology**. Technology evolves at a faster pace. Therefore, it will be considered a variable.
- **E = Execution.** Execution, although a variable from one company to another, will be assumed constant for any given company.

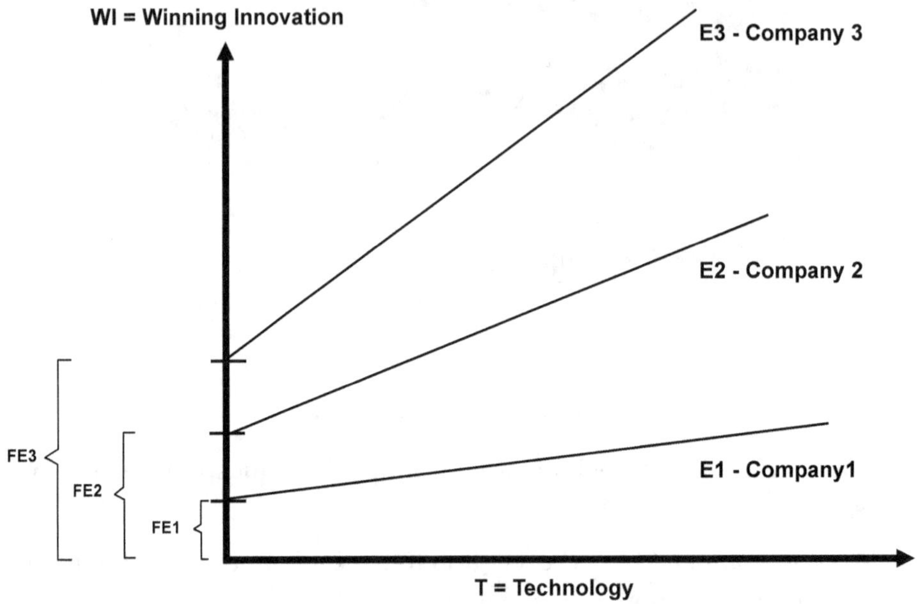

Fig. 6.6 The impact of Execution on WinnoveX™

Fig. 6.6 shows a hypothetical representation of three companies. E1, E2, and E3 in the graph above are the slopes that correspond to the three companies' execution maturity and capability levels. Assuming that E1 is less than E2, which is, in turn, less than E3, the graph in Fig. 6.6 has been created.

As conveyed by the graph, company 3 has a better preparedness to deliver winning innovations than the other two. Company 2 is better positioned than company 1, all due to the maturity of their execution capability.

Genius inventors, people who make us see things and live them before they materialize, almost like from another dimension of our beings, should not be neglected. The world will continue to appreciate their creativity. However, Equation (2) makes innovation systematic. Anybody can apply the process presented here to solve the unmet needs represented in Fig. 6.3.

Successful innovations depend on a few variables, whether relative to a product, a process, a service delivery, or even an organization. Equation (2), in its simplicity, captures them. So, let us decompose it to cover its details and better grasp its practical application.

Presented below are the various components of each element of equation (2). It is important to plan and execute them thoughtfully. For Functions Execution, it is to be noted that the functional analysis leads to the functional requirements, which are part of the overall system requirements. Technology Execution applies to all the fields of system development, including the system's (if it is a product) recovery at the end of use (conventionally known as the end of life).

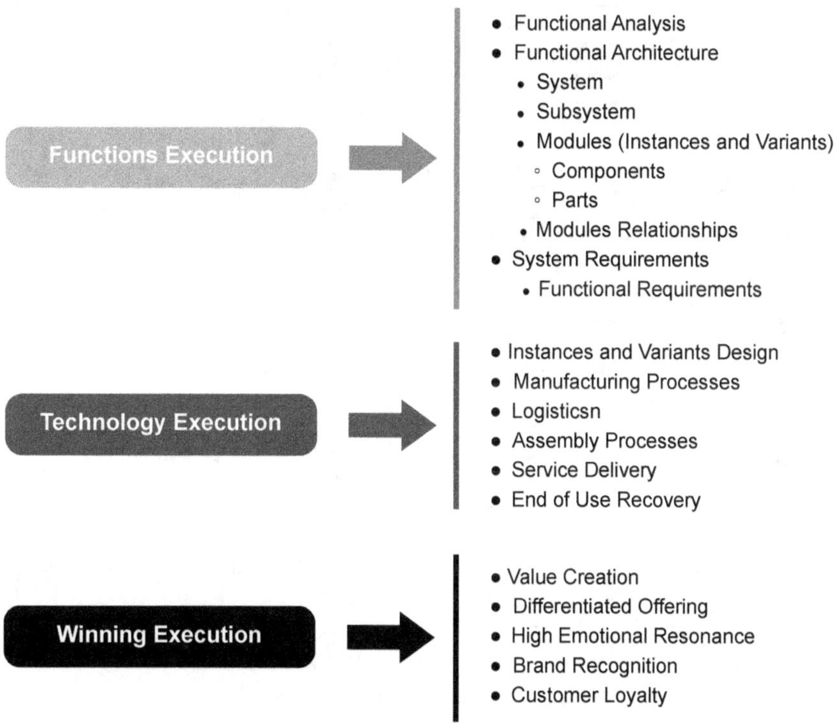

It is known in value engineering environment that:

Value = Function / Cost

The equation shows how a system's value, whether it is a product, a process, a service delivery, or an organization, is intimately linked to its function or functions. Therefore, understanding the function or set of functions of a system cannot be talked about enough. Customers buy a system's perceived value related to how much they are willing to pay to acquire the functions they want to perform. But once the function or functions are understood, the technology used to achieve them should be considered. The functions performed by a system evolve at a much slower pace than the technologies. So, they represent a suitable anchor for innovation. Technologies become obsolete. They evolve at a faster pace relative to their functions. They need to be tracked for the domains of interest and adjacent domains that could influence and even create disruption from the non-traditional competition.

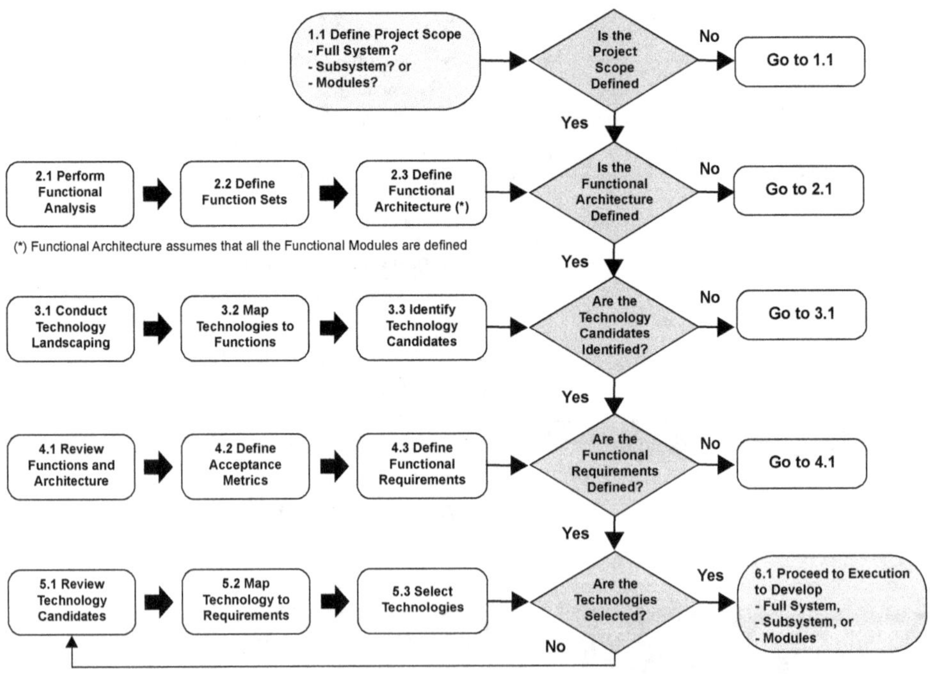

Fig. 6.7 The WinnoveX™ Way Process

The combination of functions and technologies drive innovation directly to customer value. Companies should identify the functions their systems perform to generate value to their customers. A company whose offerings are based on a Modular Product Architecture (MPA) would already have its functions library and functional architecture with all the modules defined. It would get all the benefits of an MPA and should be able to produce a steady flow of innovations within the modules for a regular system upgrade or innovate with totally new systems. The functions library is also the best driver for functional requirements since they are directly linked to customer value.

The process, described in Fig. 6.7, outlines the *WinnoveX*™ Equation implementation steps, introduced as *The WinnoveX*™ *Way*, for a journey of sustainable winning innovations.

It is important to note that technologies are considered after the functions, and the system architecture, are in place. Cool technologies that do not drive customer values should not be pursued. But if a company stumbles on a technology that is likely to generate huge customer value, no design work should start before the functions it can help perform are defined.

Let us now dive into *The WinnoveX*™ *Way* process to understand it and apply it.

- **Step 1**: Define the scope of the project. It could be a full system, a subsystem, or a module. It must be defined and documented, even if it is pulled from a long-range plan. During this step, product managers work closely with marketing and sales team members to define the scope and the market segments where the system will be later deployed.
- **Step 2**: A thorough functional analysis is performed to define the system's functional architecture. It is highly recommended that companies invest in doing this step for their existing systems. Investment in a functional architecture with a function library will pay off in the current and future initiatives.
- **Step 3**: Systems architects work with the technologists to conduct the technology landscaping and then map the technologies to the functions to identify any possible candidates. This step is an ongoing process with the companies' Innovation Management Team. Regular technology updates should be conducted with the functions relative to the unmet needs described earlier.
- **Step 4**: Led by product managers, the product development stakeholders define the requirements that need to be satisfied to meet business goals. The team will review all the available market data to determine the right metrics the functions should meet. The work is done by using the functions library. Chapter 5 describes the elements of a good set of requirements.
- **Step 5**: Once the requirements are defined, systems architects and engi-

neers map the technologies to assess the expressed needs' feasibility. They review the entire system and invite subsystem and module owners and technologists involved in the required domains to join the effort.

Steps 1 through 5 are necessary before starting any development to cover the phases of conceptual design, detail design, manufacturing and supply chain, commercialization, and release.

A few examples will help grasp the notions introduced:

Example 1: Calling an Elevator

In Chapter 2, the Operational Scenarios Review (OSR) method gave the example of passengers calling an elevator as it described how to identify a system's functions. When passengers get in front of an elevator, they push one of the floor buttons in an action known as calling the elevator. The function performed was identified as "Request Service." In most cases, requesting a service is fulfilled by pressing a button, just like in Fig. 6.8 below. However, assuming the function "Request Service" and asking a team of experts to conduct technology landscaping and match the potential technologies to the function, some exciting views can emerge.

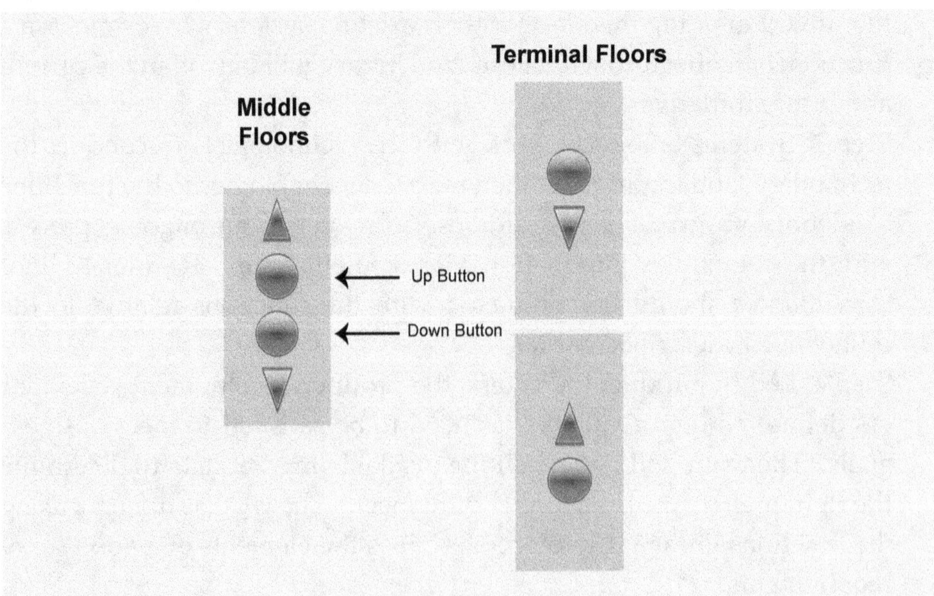

Fig. 6.8 Elevator Call Buttons

The simple brainstorming's output for technology landscaping and mapping, as explained in step 3 of Fig. 6.7, is shown in Table 6.1 below:

Function	ID	Technologies	Comments
Request Service	1	Push buttons	Existing and largely used technology
	2	Radio Frequency Identification (RFID) Key Fob—with a microchip and an antenna	Mature technoilogy from other applications such as the auto industry
	3	Pesence detection technologies	There are many mature technologies that enable the detection of a person
	4	Mobile phone app	The users of the app would be allowed to connect with the elevator controls

Table 6.1 Technology landscaping and mapping example for a module

The example of Table 6.1 is the application for a module. There were many more technologies identified, but the table only shows a few to convey the concept. When a module is considered and both its functions and potential technologies are analyzed, its interface to all the other modules it interacts with is critical and must be assessed.

Once the new module based on the selected technology is designed, the interface requirements should be met to enable its insertion into the system without disrupting its architecture. An interface can change if needed, but there should be a good case for it.

Such an exercise can be conducted for a subsystem or an entire system.

Example 2: Footwear

Modular shoe design is very trendy nowadays. The functional architecture is quite simple. The whole system is a pair of shoes. Each of them has modules that perform specific functions. The functions are:

- The primary function "Protect feet." It is decomposed into:
    - Protect sole (of feet),
    - Prevent slip,
    - Protect toes,
    - Provide cover,
    - Fit foot (each).

Fig. 6.9 represents a simple, functional view.

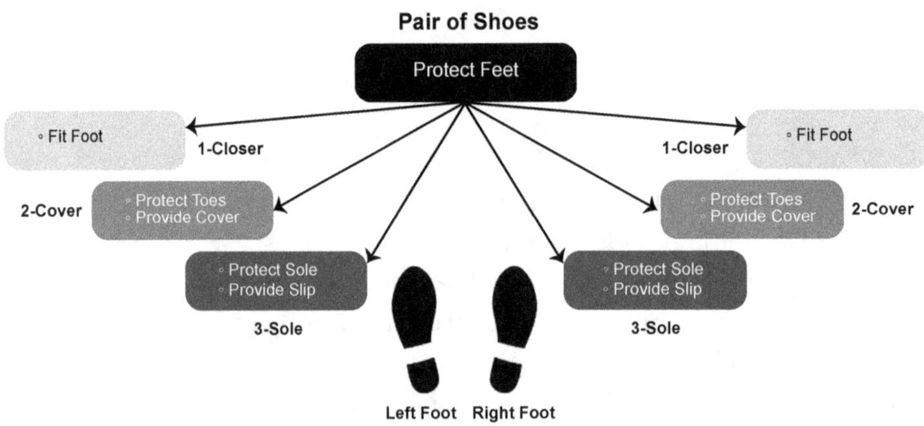

Fig. 6.9 Functions of a footwear

Then the functional architecture can be shown as in Fig. 6.10.

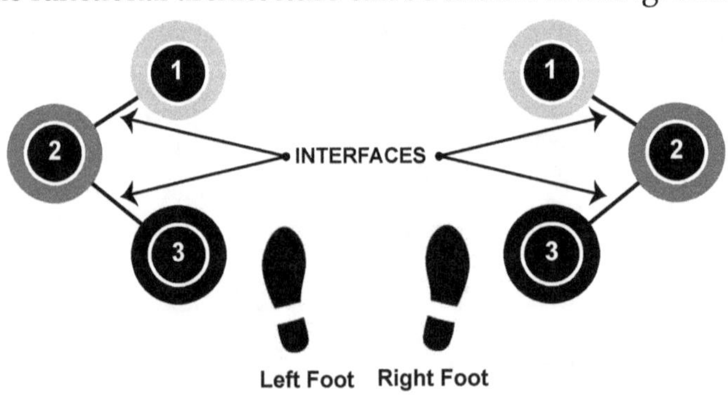

Fig. 6.10 Footwear Functional Architecture

From such a simple representation, millions of shoes instances and variants can be designed. Various technologies (materials, shapes, finishes, etc.) can be used to perform the functions. The concept has generated a fast-growing market that was not foreseen just a few years ago.

Some designers have revisited the interface between the cover and the sole of footwear, as shown in Fig. 6.11.

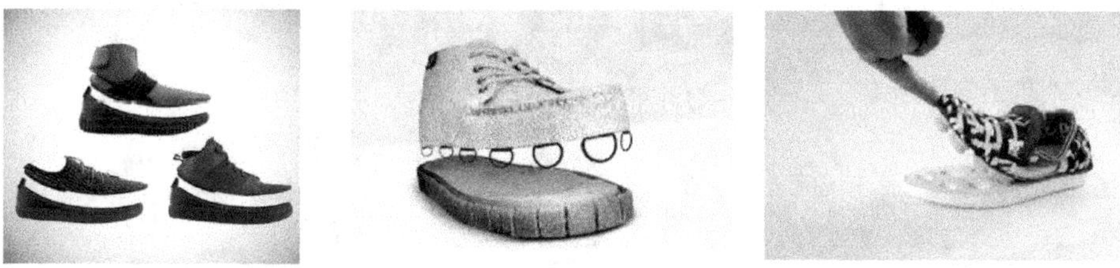

Source: ACBC.com Source: dornob.com Source: galaodesignstudio.com

Fig. 6.11 Examples of interface innovation for modular shoes

Fig. 6.11 shows the three modules (sole, cover, and closer) and their proposed interfaces in each of the examples.

Revisiting Table 2.6 of the functional subsystems and modules of a typical bicycle, countless bikes can be designed by leveraging the *WinnoveX™* Equation.

A few observations can be made as we consider Fig. 6.12, which represents a selection of bike offerings based on the functional architecture described in Chapter 2.

1. Module 1.1, representing the frame, shows many innovations based on the same functions but with different technologies and execution. Even bike 8 has the same functional architecture as all the other offerings of the figure.
2. Modules 3.1 and 3.2 (front and rear wheels) show variants driven by different technologies (bikes 4, 6, 7, and 11).
3. Bike 9, which is electric, re-uses most of a city bike 1 modules.
4. Bikes 9 and 10, respectively electric and gasoline-powered, use the same functional architecture as all the other bikes.
5. In general, the application of various technologies to the functions has been shown to generate numerous new bike designs.

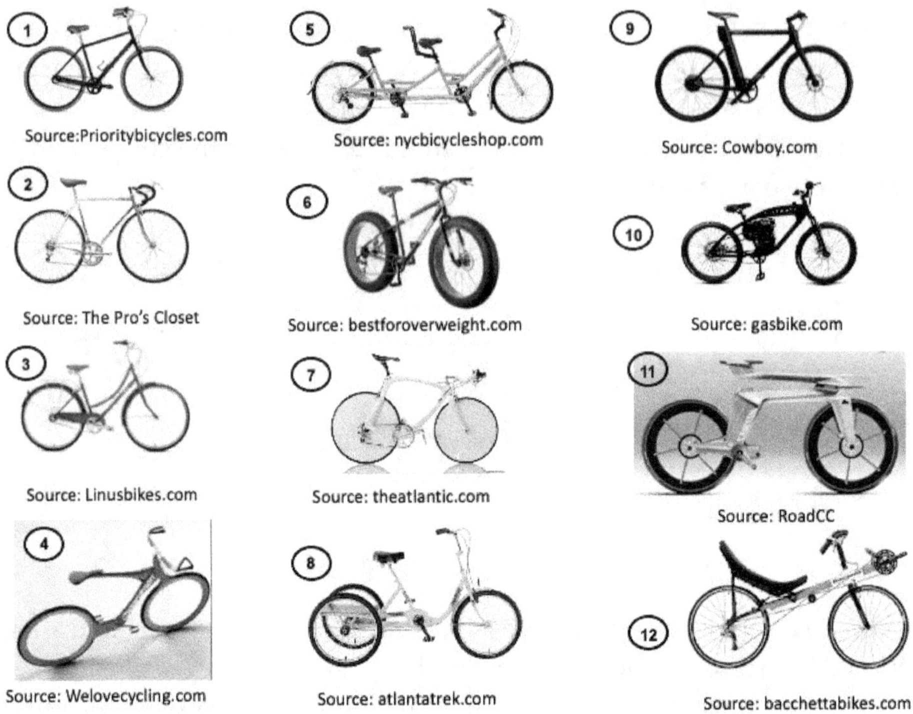

Fig. 6.12 Examples of bicycle design based on function and technology mapping

6.5 FUNCTIONS ARE THE ESSENCE

A lot was said about the value of functions, and several notions were introduced. One thing is clear. Functions are the fuel of sustained innovation and mass customization.

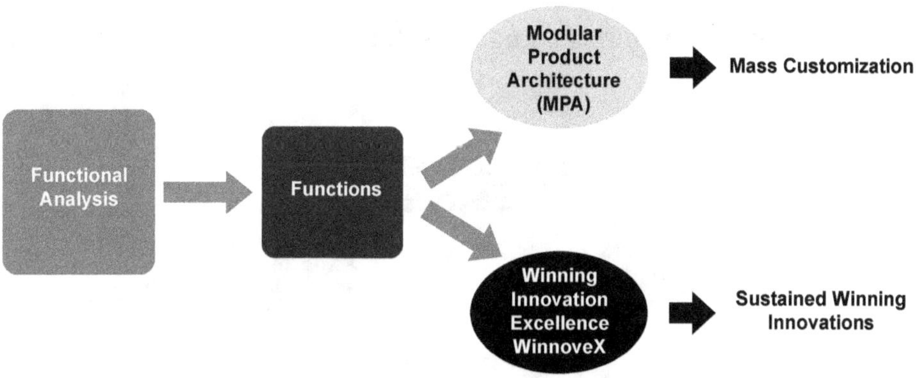

Fig. 6.13 The Value of Functions

6.6 THE WINNOVEX™ WAY SUMMARY

Innovation is the introduction of new solutions to unmet needs. When it is successful, it opens new possibilities, not previously contemplated, to all adopters. This chapter covers the types of unmet needs and shows they are not all equal and should be approached and satisfied differently. In particular, unmet needs can be known or unknown. And when an unmet need is unknown, it can be comprehensible or incomprehensible to potential customers when expressed to them. A company can then decide to address it with an incremental or a radical solution. When the various types of unmet needs and solutions are considered, six categories of addressable unmet needs emerge:

- *Known unmet* needs addressable with *incremental* solutions
- *Known unmet* needs addressable with *radical* solutions
- *Unknown, unmet* needs *comprehensible* and addressable with *incremental* solutions

- *Unknown, unmet* needs *comprehensible* and addressable with *radical* solutions
- *Unknown, unmet* needs *incomprehensible* and addressable with *incremental* solutions, and
- *Unknown, unmet* needs *incomprehensible* and addressable with *radical* solutions

The six categories of unmet needs and solutions are represented in Fig. 6.2.

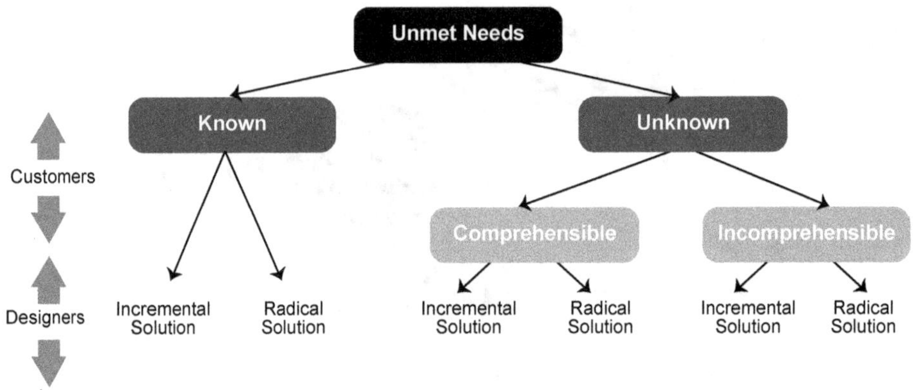

Fig. 6.2 Unmet needs categories and solution types

Each category was reviewed, highlighting how to address it with the different solution process. The analysis showed that an unknown, unmet need that is incomprehensible to customers, when addressed with a radical solution, has the highest investment risk, but it also presents potentially the highest reward. It must, therefore, be considered carefully before proceeding to next steps. On the other side of the spectrum, a known unmet need addressed with an incremental solution presents the lowest risk and is essential to keep customers engaged and assure a steady revenue stream. Evaluating the types of needs should not be a one-time exercise but an ongoing exercise aligned with product and technology roadmaps.

Innovation's fundamental equations were introduced. They highlight the critical role functions play in determining the value and driving any human-made system innovation. The two equations are at the core of the way to winning innovations or *The WinnoveX™ Way*.

The WinnoveX™ Way matches functions to technologies and drives them through good execution. Its simplest expression is through what has been defined as the Fundamental Equations of Innovation. The first equation was expressed as:

(1) Functions + Technology = Innovation

Then, adjusted with good execution it became:

(2) (Functions + Technology) x Execution = Winning Innovation

Equation (2) was simply represented by *The WinnoveX™ Way*, Fig. 6.5 below.

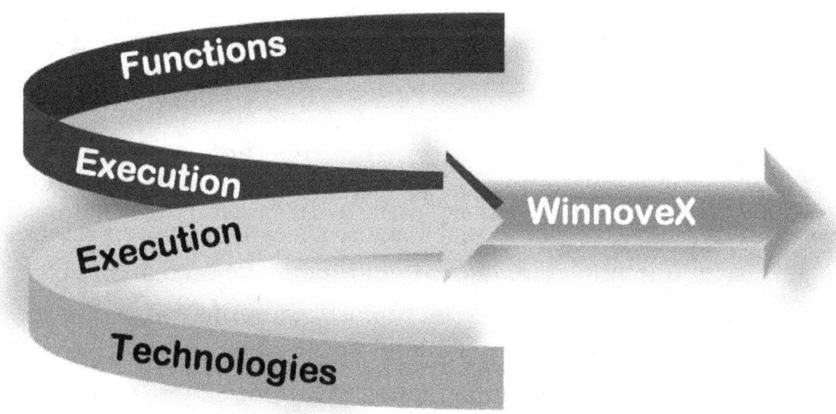

Fig. 6.5 The WinnoveX™ Way

The process that precedes any system's full development, whether it is a product, a process, a service delivery, or even an organization (as a human system), is summarized in the chart below defined as *The WinnoveX™ Way Process*.

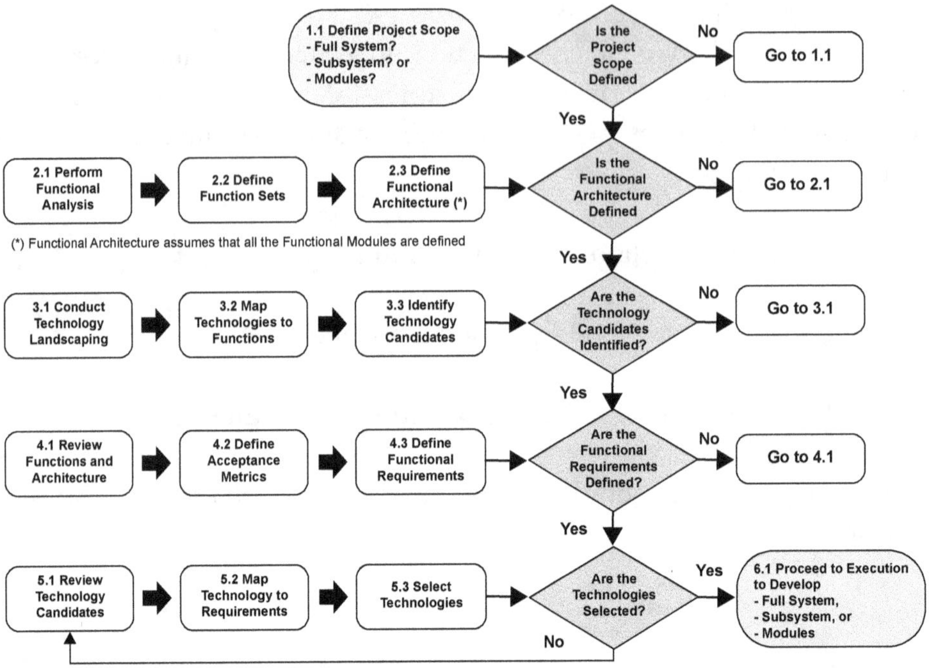

Fig. 6.7 The WinnoveX™ Way Process

It is a journey that starts with discovering the functions of existing systems (products, processes, or service delivery) or their definition for new systems through functional analysis. Functional analysis has been made simple through several methods that were presented in the previous chapters. Once the functions are defined, and the functional architecture is set, requirements can be defined. Functional requirements are the functions and the metrics that express how well they should be performed to deliver a competitive or market-leading system. With the functional requirements, technology landscaping is conducted to discover all the technologies needed to perform the functions in their best embodiments. The exercise can be conducted at a module, a subsystem, or a system level.

The best technologies are then selected and mapped to the functions. It could be just one function, a set of functions, or a whole library of functions for a subsystem or a system.

It is known in value engineering circles that there is a direct correlation between function and value. Therefore, determining the functions of a system also leads to understanding the value being offered to customers.

System development should start after the functional requirements are defined, and the right technologies mapped to them. A critical enabling asset of *The WinnoveX™ Way* is having a Modular Product Architecture (MPA). The previous chapters covered how to define an MPA. With such an asset, the field of innovation can be determined at the appropriate level (module, subsystem, or system).

Chapter 6 shows that any company can achieve innovation excellence and generate sustainable innovations along with mass customization. It also shows that by following the structured approach of *The WinnoveX™ Way* companies can:

1. Identify the functions they must perform to drive value,
2. Have clarity of vision about what they want to achieve,
3. Have the framework to perform good technology landscaping and select the solution that better performs the functions,
4. Start and sustain winning innovations, and
5. Be recognized as an innovation leader.

Part II

WinnoveX™ Applications

Introduction to Part II

Part I covered the fundamentals of *The WinnoveX™ Way*. Mainly, it showed how, for the design of a system, whether it is a product, a process, a service, or a human system such an organization, its functions are determined, and its architecture is set with its subsystems and modules. When functions are mapped to the right technologies with good execution, the process generates winning innovations. The methods presented were validated over many years of application in the operations beyond the design phase. Part I also showed how a Modular Product Architecture (MPA) enables and enhances the innovation methods presented.

Part II will explore the operations spectrum to present the findings and open the door to innovating in any company that decides to travel along *The WinnoveX™ Way*.

Fig. II-1 The WinnoveX™ Way Application Areas

In Part II, MPA is thoroughly woven into *WinnoveX*™ applications.

Fig. II-1 summarizes the spectrum of all the areas where the *WinnoveX*™ Equations apply.

Beyond the Design phase mostly covered in Part I, Part II will cover the following areas:

- Manufacturing and Supply Chain
- Logistics (covering packing, loading, shipping, and transportation)
- Installation
- Aftermarket and Service
- Product Management

Also, due to their importance, a couple of domains will be added, namely:

- Modeling & Simulation
- Organizations Design and Human Dynamics

Part II will end with a few words on some technologies that have the potential to shape the 21st Century.

- A Few 21st Century Technologies

7

Manufacturing and Supply Chain

7.1 MANUFACTURING OF MODULAR SYSTEMS

Any physical system's materialization goes through the vital step of manufacturing, where the system's elements are produced and assembled. Like any combination of art and science, manufacturing has evolved with the introduction of various technologies in the course of history. Let us now explore how a modular product architecture transforms an assembly process.

On conventional manufacturing floors, it is common to notice the movement of parts, components, or modules on and across assembly lines. Some system elements have quite a travel from start to finish of their processes. The introduction of the so-called spaghetti diagrams often shows the long distances covered on the factory floor by items and people as the material being worked on and assembled moves during production. By plotting a product's subsystems' flow lines in a conventional factory, they show components and modules moving from one station to another, often crossing paths before finishing at a common place for their assembly. Such a discovery led to the study of the phenomenon, and several lean models were proposed. Still, when multiple product variants have to be produced in the same factory, leading to various configurations' assembly on the same floor, the production flow optimization is not a simple task. However, in a modular product environment, the architecture allows the implementation of a few features:

- The creation of subsystem lines and single flows for the manufacturing and assembly of the modules.

- The collocation of parts and components along module production lines for better visual management.
- A better collocation of production machines covering several subsystems, especially machines that are expensive and cannot be placed on only one line.
- Ease of implementation of in-line and end of line testers.
- A common area for production materials reception is mostly collocated at the production lines' front end to feed them.
- A common area for subsystem consolidation at the back end of the production lines.
- A common area for systems consolidation by contract and ease of picking to load the vehicles.
- Overall a better layout for the floor.

All the factory floor features identified lead to a consistent layout that can be replicated at different locations.

A few models of floor layouts that benefit from modular products are presented next. In particular:

- Assembly lines of small systems,
- Assembly lines of medium and large "plug and play" systems,
- Factory lines of systems assembled at the job site (after manufacturing)

Modular products offer the opportunity to simplify factory lines and improve material flow on the floor.

7.1.1 Small Systems Assembly Lines

For small systems with few modules such as pushed lawnmowers, it is advisable to pre-assemble each of the modules and deliver them at the right station through the material feeders, as shown in Fig. 7.1. The module interfaces should be made simple. For example, they could consist of bolted solutions, snap-ons, and or connectors. Modules would be progressively integrated from station 1 to station 4; then, the full system assembly would be at station 5 before the end of line testing station.

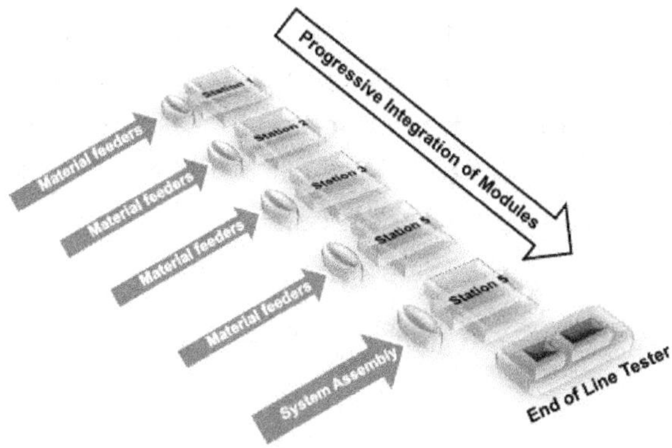

Fig. 7.1 Module-based small systems assembly line

7.1.2 Medium to Large Systems Assembly Lines

For assembly lines involving complex in-line operations of large subsystems, such as welding by robots or other automation, a subsystem may have one or more lines. Still, it is advisable to manufacture and pre-assemble the modules' components as much as possible in other work cells outside of the subsystem assembly lines. The model proposed in Fig. 7.1 shows the main material receiving and inspection area at the front end of the lines. Each station has feeders. At the back end of all the lines, the final assembly area is reserved. The assembly area could also, due to other factors, be located somewhere other than the back end of the subsystem line. In such cases, a materials movement simulation is advised to minimize the distances mentioned earlier.

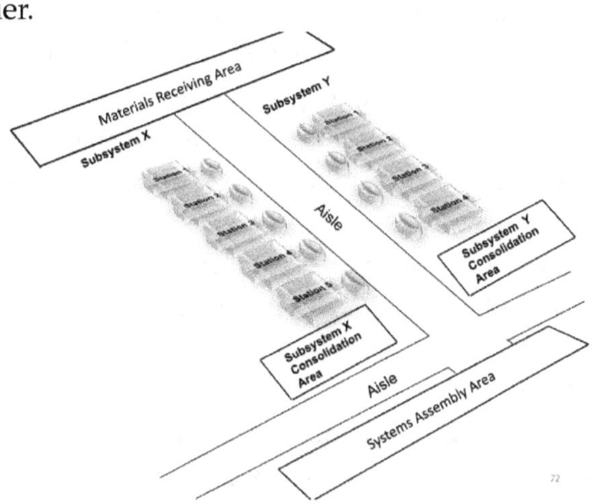

Fig. 7.2A Module-based factory floor for fully factory assembled systems

Some subsystems made of complex modules could have several lines with each line for a different module or group of modules. If a specific module line requires a limited number of stations and is short, another module line of the same subsystem could be added for better space utilization. Such an arrangement is represented by Fig. 7.2B.

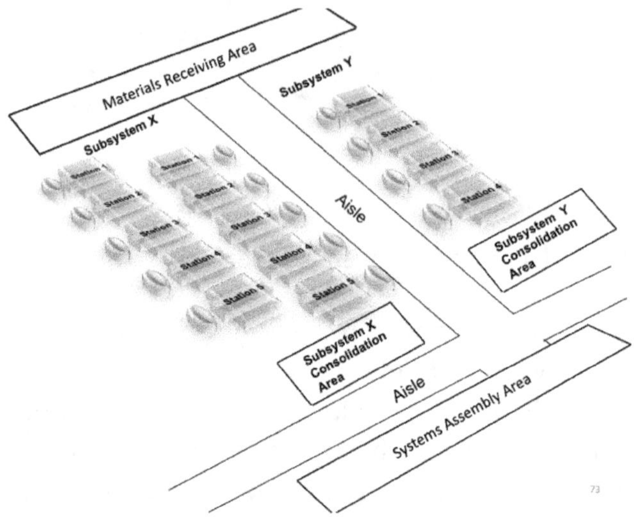

Fig. 7.2B Module-based factory floor for fully factory assembled systems – with multi-line subsystems

7.1.3 Medium to Large Jobsite Installed Systems Assembly Lines

Some systems such as elevators or HVAC systems connected to ducts and other air movers need to have their subsystems delivered and installed at a jobsite. Those are not "plug and play" systems fully assembled from a factory. After manufacturing, consolidating, and transporting all the subsystems to the jobsite, the final assembly requires skilled installers to make all the connections before the system is powered, commissioned, and released to the customer. In this case, the proposed model in Fig. 7.3 is slightly different from the one presented in Fig. 7.2. A picking area where the subsystems are loaded and transported to either a logistic center or a jobsite would replace the final assembly area.

The figures above show how a module-based factory floor can be modeled and consistently replicated at all the production facilities of a portfolio of products that share the same functional architecture, even if the instances and their variants are significant.

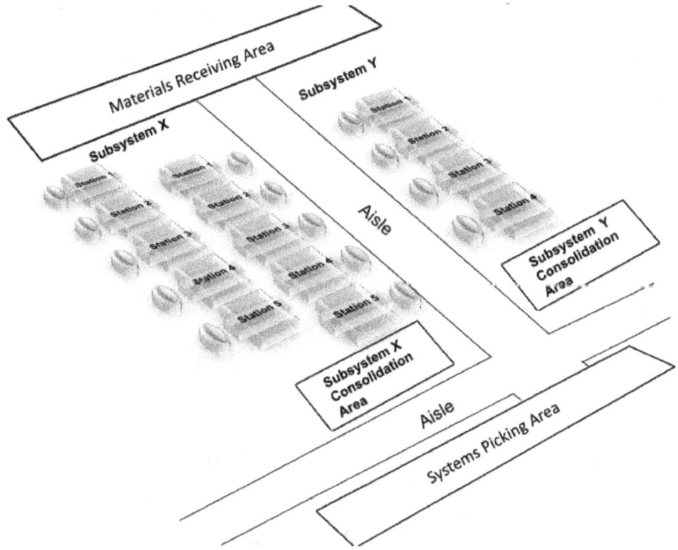

Fig. 7.3 Module-based factory floor layout for jobsite installed systems

Even though only two generic subsystems, X and Y, are shown in Figures 7.2 & 7.3, several subsystems would be put side by side between the materials reception area and the final assembly or subsystems consolidation areas at the end of the lines.

In Part I, it was mentioned that some non-core modules could be outsourced with their design and production coming from suppliers. Core and non-core modules could also be integrated and managed by Subsystem Supply Partners (SSP). When a decision is made to engage an SSP, the company must prepare the right documents to ensure the subsystems' modules have the interfaces needed to secure the whole system's integrity. For systems that are fully assembled in the factory, the SSP's subsystems should be delivered directly at the final assembly area or the subsystem consolidation area. Some companies create a logistic center for systems installed at customer sites to consolidate full system packages before delivery. In such a case, with well-architected systems, the boxes coming from an SSP could be delivered at the Systems Logistics Centers (SLC). SLC's are also used to keep systems whose installation, for some reason, has been delayed.

7.2 SUPPLY CHAIN

The different stages of value delivery in a company involve internal and external resources. The choice of the external resources that make up the links of the chain needed to meet the company's business and financial goals is critical to the attainment of objectives related to:

- Capital investment in equipment and other assets
- Product quality and reliability
- On-time delivery
- Inventory turns
- Capacity utilization,
- Etc.

Some companies choose to vertically integrate to better control some or all of the objectives mentioned above. The vertical integration aspect will not be covered in this book to explore the new opportunities created by modular products or systems environment. We would rather spend time understanding what a modular system architecture means to a company, whether it is a finished-system or a subsystem, or even a module provider.

Every manufacturing company is continuously reviewing and optimizing its processes to operate with negative working capital, short lead time, higher inventory turnover, etc. To reach such vital goals, several companies strategically choose to Make to Stock (MTS), Assemble to Order (ATO), Make to Order (MTO), or Engineer to Order (ETO) the products they deliver. Small items, close to commodities, can be made to stock, but this strategy is not used for medium and more massive systems.

Even complex subsystems delivered for integration into large systems often need to follow either an MTO or an ETO route. But with the trend to provide mass customization, the pressure is even higher on the companies' product development teams to consider it and all the other already challenging goals. Unfortunately, some companies want to mass customize without the fundamental ingredients. Therefore, we will tackle, in the next session, how product modularity helps these strategic choices. We will also review how the material cost of a product could be structured in a Modular Product Architecture (MPA) environment.

7.2.1 Manufacturing Strategic Choices

As described above, companies must make some strategic manufacturing choices to meet their market demand and their business objectives.

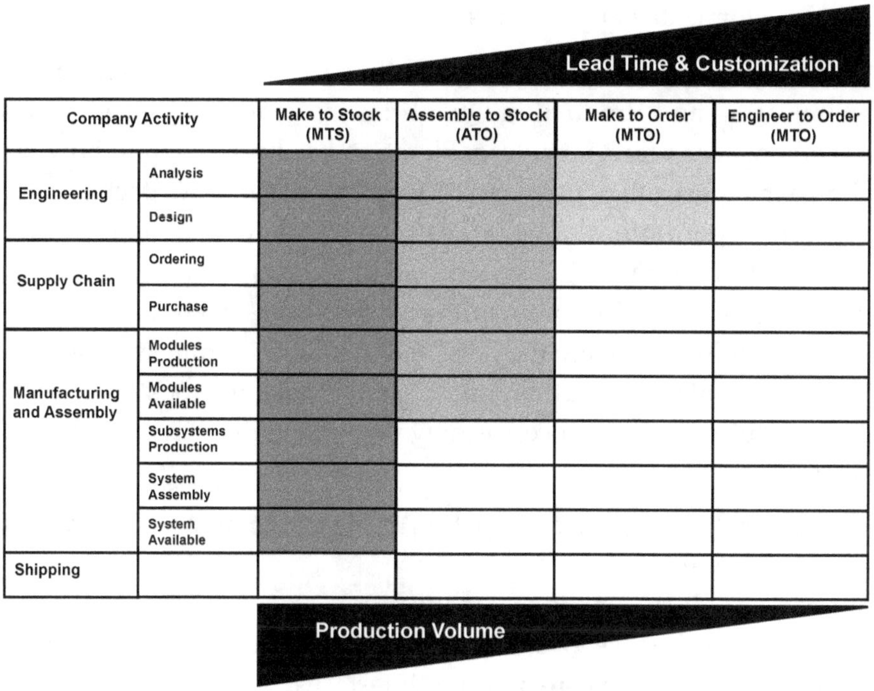

Fig. 7.4 Manufacturing Strategic Processes

The well-known options are summarized in Fig. 7.4.

- **Make to Stock (MTS).** Finished goods are made and stored, ready to be delivered. It offers a short lead-time and is suitable for systems that are not complex and relatively inexpensive to avoid the stagnation of precious financial resources since they can be produced in large volumes.
- **Assemble to Order (ASO).** Engineering and sub-assemblies manufacturing are done. The components are stored and assembled when an order arrives. The option is also suitable for a short lead-time, but it has little room for customization.
- **Make to Order (MTO).** Engineering is done, but nothing is manufactured until orders are received. It allows more customization but requires a longer lead-time than MTO and ASO.

- **Engineer to Order (ETO)**. The trigger, in this case, is the customer's order. Engineering, manufacturing, and assembly are all done once the order is received. While it allows for maximum customization, it has a long lead time and is applied to complex systems with low volumes or highly customizable applications.

When the processes are put side by side, the trade-off between Lead-time & Customization and Production Quantities influences the level of mass customization. Therefore, some consideration has been given to how best to optimize the spectrum of processes. Modular Product Architecture offers some unique opportunities. The key to success is having a modular architecture with clearly defined subsystems and modules with their interfaces. Customer requirements are met with predefined, designed, and manufactured core modules. They are the ones that would determine the lead-time due to their complexity, but their anticipated "make to stock" cuts down lengthy schedules. Such a process is only possible with analysis and investments in:

- Deep understanding and anticipation of the conscious and unconscious market unmet needs,
- Company products functional architecture,
- Clearly defined subsystems and modules with their instances and variants and control of the core modules interfaces,
- Root BOM defined and leveraged for manufacturing cost, logistics, and installation,
- Establishment of Subsystem and Module Supply Partners,
- System fully modeled with customer order simulation to accelerate the specification review and design approval.

With such investments, an ETO's flexibility can be matched with the lead-time and production volume of an Assemble to Order (ATO). Some manufacturing experts talk about Configure to Order (CTO). However, what this section describes, which has been implemented by a few companies, goes beyond the configuration. Design specifications, cost estimates, proposal review, and approval are all based on the anticipated needs analysis and system simulation capabilities combined with the design and even the pre-production of the core modules. The process is defined as *Winnovate to Needs* (WTN), as shown in Fig. 7.5. A Modular Product Architecture simplifies implementing an Assemble to Order (ATO). It enables the same flexibility and lead-time for more complex systems through a WTN adoption.

Fig. 7.5 shows how lead-time & customization and production volume could be the same for ATO and WTN. Yet, in the chart, only WTN presents the opportunity to offer more complex systems.

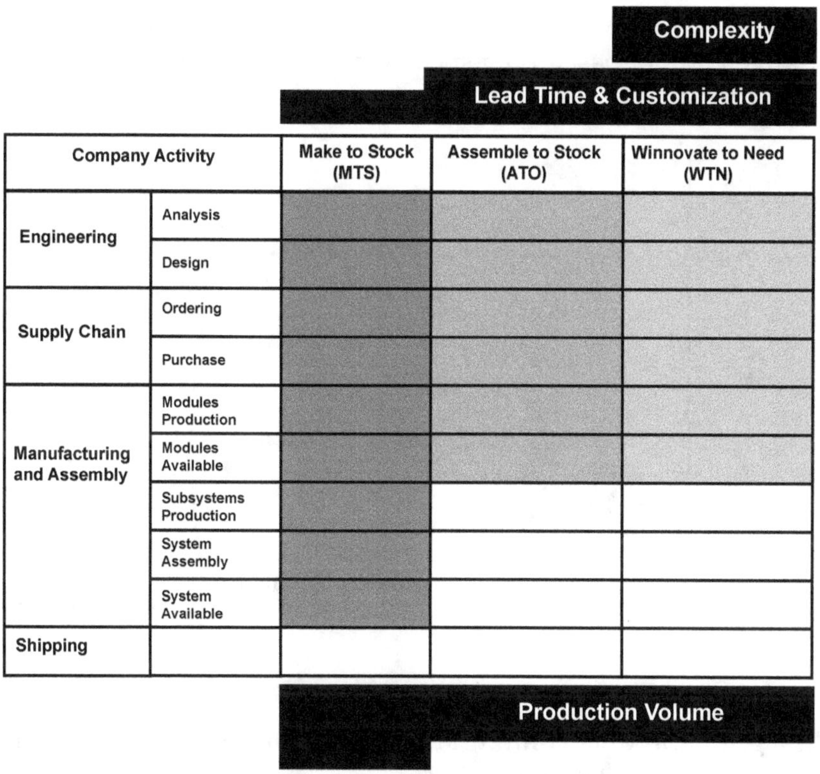

Fig. 7.5 Winnovate to Needs (WTN)

In general, cost is one of the most critical parameters to control for any business. It is even more so in a production environment. Therefore, the next section describes the opportunity that modular products present to material cost management in a factory.

7.2.2 Product Factory Material Cost

As presented in Part I, the BOM is also based on subsystems and modules in a modular product environment. Product material cost is covered in this chapter, and the installation cost will be reviewed in another chapter.

Once the product portfolio's functional architecture is set, it is important to leverage it for the cost structure for which two approaches will be reviewed next.

7.2.2.1 The Reference Product Configurations Cost (RPCC)

For each of the products that make up the portfolio of a company's offerings, some configurations are often the most sold. Fig. 7.6, which is only illustrative and not based on a specific industry, represents sales volumes' data-driven observation.

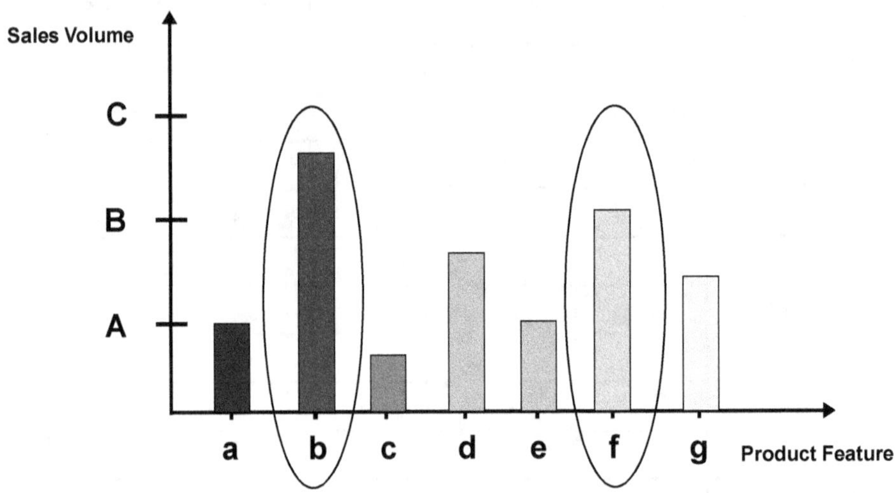

Fig. 7.6 Illustrative representation of sales volumes for a product feature

In Fig. 7.6, product configurations with features b and f are the most sold. A feature would be a metric for a specific product type. For example, it could be kilowatts (kW) for rotary air compressors, load capacity in kilograms (kg) for passenger elevators, or wheel size in millimeters (mm) for road bikes. Once these high-volume units with their features are identified based on their module instances and variants, they can be selected as reference units. There could be more than one reference unit. The number of references depends on the portfolio size and the spread of the units' cost. It should be set to represent the portfolio. Each reference unit's detailed cost is then prepared and is used to estimate the cost of the other units, it represents. Table 7.1 below is an example of a template that could be used. This case is from a road bicycle from a bike portfolio that could include several sizes of road bikes, mountain bikes, race bikes, motorized bikes, electric bikes, etc. A standard cost matrix is established for each of the reference units. Once the module variants' cost is compiled, each subsystem cost is computed, followed by the whole system (bike) cost. Suppose the same product is manufactured in several locations. In that case, a comparison becomes possible, and supply chain anomalies can be resolved to always move toward the most competitive global cost for similar requirements.

SS ID	Subsystems	Mod. ID	Modules	Modules Instances/ Variants #	Components	Unit Cost	Quantity	Mo
1	Frameset	1.1.	Frame					
		1.2	Fork					
		1.3	Fender					
2	Propulsion	2.1	Pedals					
		2.2	Motion transmission					
		2.3	Speed selector					
		2.4	Motion converter					
		2.5	Power converter					
3	Wheels	3.1	Front wheel					
		3.2	Rear wheel					
4	Bike Control	4.1	Steering assembly					
		4.2	Acceleration lever					
		4.3	Stopper					
5	Brakes Assembly	5.1	Front brake					
		5.2	Rear brake					
		5.3	Brake cable					
		5.4	Brake levers					
6	Dashboard	6.1	Speedometer					
		6.2	Charge indicator					
7	Load Carrier	7.1	Seat					
		7.2	Seatpost					
		7.3	Luggage Carrier					
		8.1	Generator					
		8.2	Engine					

Table 7.1 Bike reference material cost model (based on a standard cost matrix driven by the RBOM)

ΔP Full table is in Section D1, Appendix D.

The cost considered here is essentially the material cost. It is the cost of each module at the assembly line. To get each subsystem's total cost, premier companies capture each module assembly time and then convert it into cost, knowing the cost of every minute on the line (factory's time unit cost). The costing template shown in Table 7.2 is a powerful reference when the same system is produced in several locations. A cost comparison is always in action to share best practices and continuously converge toward the most competitive assembly time/cost.

SS ID	Subsystems	Mod. ID	Modules	Components	Module Material Cost	Module Assembly Labor Cost	Total Module Cost
1	Frameset	1.1.	Frame	Frame, Rear Shock absorber			
		1.2	Fork	Fork, Bearings, Nuts, Front shock absorber			
		1.3	Fender	Front fender, Rear fender			
2	Propulsion	2.1	Pedals	Right pedal, Left pedal			
		2.2	Motion transmission	Chainrings, Crankarms, Spider, Chain			
		2.3	Speed selector	Cassette, Front derailleur, Rear derailleur			
		2.4	Motion converter	Crankset			
		2.5	Power converter	Drive			
3	Wheels	3.1	Front wheel	Rim, Hub, Spokes, Bearings, Tube, Tire			
		3.2	Rear wheel	Rim, hub, spokes, bearings, tube, tire			
4	Bike Control	4.1	Steering assembly	Handlebar, Headset, Stem, Grips			
		4.2	Acceleration lever	Acceleration Lever/Shifter			
		4.3	Stopper	Stopper			
5	Brakes Assembly	5.1	Front brake	Disc, Caliper, Pad			
		5.2	Rear brake	Disc, Caliper, Pad			
		5.3	Brake cable	Front cable, Rear Cable			
		5.4	Brake levers	Front lever, Rear Lever			
6	Dashboard	6.1	Speedometer	Speedometer, Wire, Wheel hub connector			
		6.2	Charge indicator	Display, Wire, Battery clamp			
7	Load Carrier	7.1	Seat	Saddle, Saddle clamp			
		7.2	Seatpost	Seatpost, Seat clamp			
		7.3	Luggage Carrier	Luggage Carrier			
8	Power Generation	8.1	Generator	Generator, Generator clamp			
		8.2	Engine	Engine, Clutch, Exhaust pipe,			
		8.3	Tank	Container, Filter, Cap, Hose, Tank clamps			
		8.4	Fuel	Fuel			
		8.5	Electric motor				
9	Energy Storage	9.1	Battery pack	Battery, Battery clamp, Charge electronics			
		9.2	Charge supervisor (SW)	State of charge software			
10	Lights	10.1	Headlight	Headlight, Headlight clamp, Wire			
		10.2	Rear light	Rear light, Rear light cla			
			Total System Cost				

Table 7.2 Total Module and Subsystem Cost (including assembly cost)

ΔP Full table is in Section D1, Appendix D.

7.2.2.2 Product Portfolio Cost Database Method

The second cost document drives the creation of a cost database that covers all the variants of module instances, followed by a selection tool. The tool is then linked to a configurator that automatically draws all the selected items to make the system cost. The configurator allows the discrete selection of all customer choices then computes the cost. It can also be programmed to create all the order data sheets and send them to internal manufacturing and the vendors with no further intervention.

The cost of all the products manufactured by a company should be known by the internal stakeholders. The company should develop a way to quickly estimate the modules' cost during their development and then track it to the end of the product release. Some excellent software packages on the market help perform the "should cost" analysis of a product's modules during their development. Companies should not limit themselves to the practice of sending drawings to a vendor to get quotes and compile them to get a product's final cost. Although such an approach is still in use worldwide, it may be time for all product development teams to develop the capability to estimate the cost of their products and compile the actual cost accurately before the product release. Any cost provided by Subsystem Supply Partners, or by any supplier, due to supply contracts, should be negotiated based on internal estimates before entering them into the database.

7.3 THE FINAL PRODUCT AND SERVICE TREE

The concept presented here is the tree made of all the suppliers in a company's value chain. The structure describes the links in a company's value delivery chain, which most companies do not know. It also shows a company its level and position in a value chain. It may be too hard to know all of them, especially for large companies, but it helps to know more than two levels.

While we discuss the general application of Final Product Tree or FPT, one specific example of its application is the Carbon Footprint assessment for a company's products. Many companies are vowing to reduce their CO2 emissions, but how can they succeed if they do not have the landscape of what contributes to their products or services along their value chain? The Carbon Footprint reduction will not be a subject of this book. Our focus instead would be on how to maximize value by leveraging a Final Product Tree (FPT).

Within the final product tree, companies that provide their products or services to the consumers in a B2C relationship face different issues and opportunities than those whose products or services are integrated into other products in a B2B relationship. The following figures present several B2B and B2C scenarios to highlight some supply chain opportunities that may be hiding in plain sight.

For any company, the ideal state in a B2C relationship is to deliver value to the customers while minimizing the investment in physical assets regarding their acquisition and maintenance. It is even better if most of the physical assets are owned by partners while the company is recognized as the value creation and delivery brand. The model is represented in Fig. 7.7.

Fig. 7.7 B2C relationship with minimum assets

Some 21st Century companies (Uber and Airbnb, for example) have seen great success with such a model. Their business model is based on an added value in the direct connection of available space with customers in need. Such companies manage customer interactions and provide products and services through software and own none of the physical assets.

However, for typical companies that make physical goods, there would be a line of providers that supply the sub-assemblies they need to fulfill their mission and deliver products to their consumers. Such a company's relationship with its first line of providers would be a B2B relationship, even though the company may still have a B2C relationship with its customers. The two types of relationships are synthesized in the model expressed in Fig. 7.8 showing suppliers' first line as Final Product less One, or FP-1.

Fig. 7.8 B2C and B2B Model showing the first line of suppliers

The concept can be generalized with more layers behind the company serving the consumers. The structure forms the final product or service tree. It also highlights the B2B relationship between all the support layers. Fig. 7.9 shows multiple layers of suppliers.

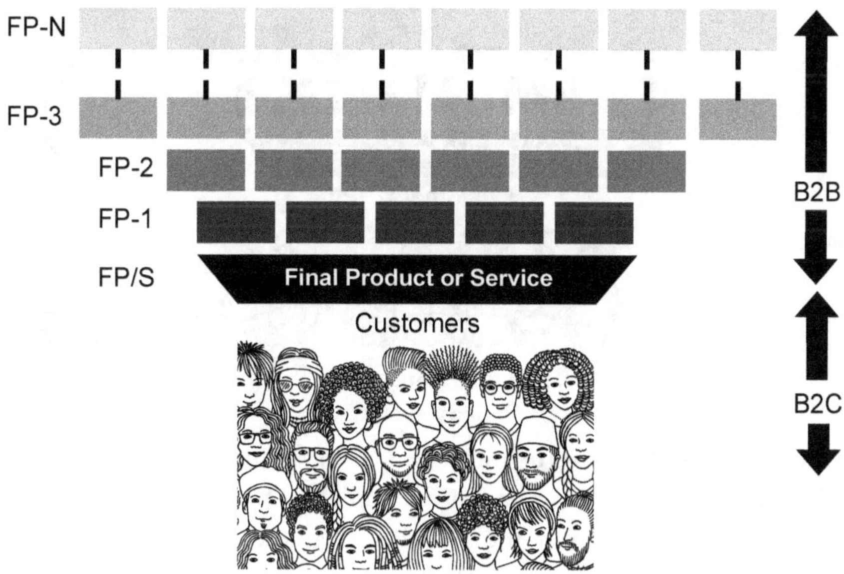

Fig. 7.9 B2B with multi-layers suppliers

For final product or service companies operating in a modular system environment, it is highly beneficial to establish and deal with Subsystem Supply Partners (SSP). In case some modules are complex and would be better designed and manufactured by companies specialized in the required technologies, it is beneficial to have Module Supply Partners (MSP), as illustrated in Fig. 7.10. They would take on the responsibility of integrating entire subsystems or special modules and deliver them for full assembly. The final product or service provider still has the responsibility of the final assembly. A strategic decision to consider is to oversee the design and manufacturing of the core modules or have them produced or even designed and produced by an SSP. The introduction of SSP's simplifies the product delivery tree and raises the accountability of the suppliers. Once any company invests in understanding and drawing its full product delivery tree, it can architect and simplify its supply chain with great benefits. Strategic planning can be done with the SSP's and the MSP's that could lead to the development of technology financed by their investments.

Fig. 7.10 Illustration of a Preferred Subsystem Supplier with its layers

For companies that are further away from the final product or service provider (see Fig. 7.11), it is harder to achieve strong bargaining power. Unless they specialize in extremely critical technologies with limited to no competitors, they will be considered commodity providers and squeezed for cost. It is not an enviable position. However, for companies already in such a place, they must perfect their craft and differentiate from their direct competitors.

Fig. 7.11 B2B Supplier further away from the final product or service provider

7.4 WINNOVEX™ IN MANUFACTURING & SUPPLY CHAIN

In the previous sections of this chapter, we reviewed modular products' impact on the organization and factory operations management. In their book entitled Factory Physics, Wallace J. Hopp and Mark L. Spearman explain what "factory physics" is and why it should be studied. Their short answer is that "factory physics is a systematic description of the underlying behavior of manufacturing systems."[1] Just as they recognized the need and covered the foundations of manufacturing management, here too, we highlight how the equations introduced in Chapter 6 of this book can be used to drive winning innovation excellence in manufacturing and supply chain.

A factory that is set up to manufacture modular products can, itself, undergo the analysis of the functions it needs to perform to deliver as expected. The same system analysis approach applied to existing and new systems can also applies to existing and new factories. The functions to be performed to deliver the expected output can be set and allocated to the various sections of the production flow through decomposition and grouping. Then, just like it is done to design any system, the right technologies should be researched, selected, and applied to determine which hardware each area and line need to meet their requirements. So, here again, the same fundamental equations of innovation from Section 6.4 apply:

(1) Functions + Technology = Innovation

(2) (Functions + Technology) x Execution = Winning Innovation

Let's assume, for example, a factory specialized in the manufacturing of cabinets for controls equipment. First, the product requirements would be set, and they would determine the design and acceptance criteria of the control cabinets. The functions the factory equipment needs to perform to satisfy the cabinets and the factory's performance requirements will be identified and mapped to selected technologies to deliver the expected quality, cost, output quantities, minimal investment in assets, and other needs of the

1 *Factory Physics*, 2nd Edition, Irwin McGraw-Hill 2000, p.1

factory to give it a competitive advantage. Lastly, the equipment with the right technologies able to perform the functions identified would be selected to populate the factory's lines. Here also, by establishing the functions of the production flow (equipment and processes), clarity and innovation are brought to the design and implementation of any manufacturing and supply chain operation.

7.5 MANUFACTURING AND SUPPLY CHAIN SUMMARY

Any physical system's materialization goes through the important step of manufacturing where the system's elements are produced and assembled. In this chapter, we have reviewed some concepts that are aided by modular product architecture implementation. Assembly lines can be simplified on a factory floor. MPA drives the floor layout standardization, which all the plants that produce the same offerings can replicate across the world. A couple of product cost models were presented. They could be adopted and adapted to better manage the material and assembly cost in companies that integrate their products with modules and subsystems from various sources.

Some emphasis was put on the different stages of value delivery in a company that involves its own internal and external resources. The choice of external resources to support the company's business and financial goals is critical. Good choices lead to capital investment in equipment and other assets, cost optimization, product quality, reliability, on-time delivery, inventory turns, capacity optimization, space utilization, etc.

The notion of a final product or service tree was introduced to model the various layers of suppliers involved in the delivery of a final product or service. In a modular system environment, the supply chain can be simplified by engaging Subsystem Supply Partners (SSP). They can assume the responsibility of the manufacturing and even the design, manufacturing and financing of entire subsystems. If some product modules are critical and better managed by specialized suppliers at a competitive cost due to their domain knowledge, Module Supply Partners (MSP) can also be engaged. Both SSP and MSP should work with the company on strategic decisions, including investments to deliver their subsystems or modules of responsibility for the product's final assembly. Any company in the supply chain can assess its position on the Final Product Tree and adjust its strategy accordingly. For example, a company in a hidden layer from the final product should differentiate itself from its competitors to avoid being a commodity provider with the risk of being forced to compete at lower prices.

Modular Product Architecture enables the application of *WinnoveX*™ equations and helps manage:

- Cost and assembly time,
- Factory floor layout, product flow, capacity optimization, and
- The Final Product Tree to improve quality, on-time delivery, inventory turns.

The application of *The WinnoveX*™ *Way* steps leads to identify the functions that must be performed by the factory equipment with the right technologies to deliver the product while making the factory efficient, competitive, and profitable.

8

Logistics

This chapter intends to highlight the strategic nature of logistics, especially when taking full advantage of a modular product architecture during the journey on *The WinnoveX™ Way*.

Initially applied to the management of military supplies in the 19th Century, the term logistics has found great relevance in the business domain. It can simply be defined as the management of the flow of equipment and information between suppliers and manufacturers and between manufacturers and consumers. Therefore, there are two flows to consider, as shown in Fig. 8.1. The first flow between suppliers and manufacturers is also known as the "Inbound" flow, and the second between the manufacturers and the consumers, the "Outbound" flow.

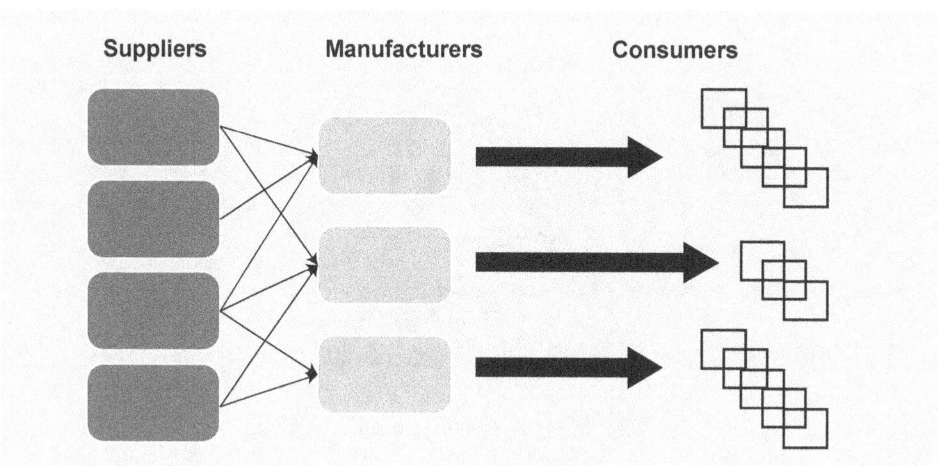

Fig. 8.1 Flow from suppliers to manufacturers (Inbound) and flow to consumers (Outbound)

Both flows have in common:

- Packaging/Packing[1]
- Material handling
- Transportation
- Warehousing
- Quality
- Time
- Dependability
- Security, and
- Information management

These common factors can be broken into quantitative metrics and qualitative requirements for the assessment of their goodness. Logistics can be and are modeled to better manage what is known as Total Logistics Cost vs. any unanticipated cost such as damage during transportation.

Such a study enables a lot of trade-offs. For example, the fastest means of transportation may be the most expensive and may also have more constraints about packages' size and weight, but it may be the most dependable, secure, and with the least damages. For a bulky load with tolerance toward lead time, another means of transportation may be better. My personal experience leads me to consider finding the best balance between quality, cost, and time, which most experts in logistics chase all day.

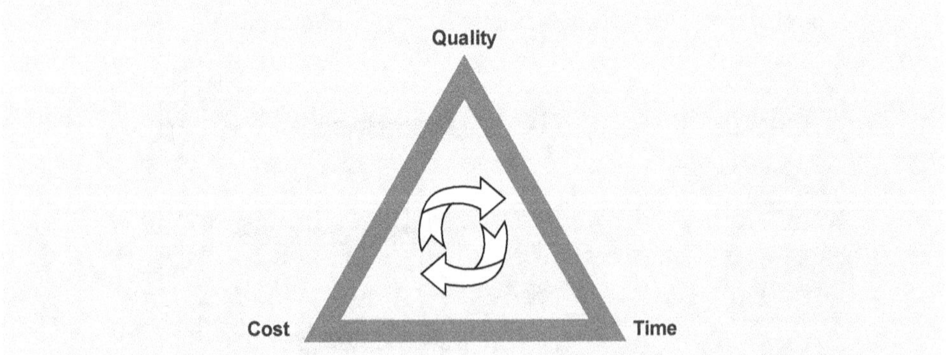

Fig. 8.2 Balancing Quality, Cost, and Time

1 Even though there is a difference between packing and packaging, the two terms are used here indifferently to express both the protection of a package as well as the process of making the protective material.

8.1 PACKAGING[2] BILL OF MATERIALS VS. SHIPPING BILL OF MATERIALS

The two notions that will be introduced here for use in the next sections are:

- Packaging Bill of Materials or PBOM: A list should express any package's content enabling its tracking from one location to another and its inspection at delivery. It defines the PBOM, only relative to the package, and can be either inserted inside or put on an external label with a readable tag. It can also be expressed through both an inserted list and a label.
- Shipping Bill of Materials or SBOM: When several packages are grouped into a larger container, its content should also be described through a list that specifies each of them without getting into their explicit content. It is defined as the SBOM. For a modular product, the SBOM is essentially the list that identifies the subsystems or cross-modules grouped in a single container.

The PBOM reflects the products Root BOM and should be in line with the subsystems and their modules and, in some cases, their instances and variants for more clarity. The nomenclature of the SBOM is usually defined by the manufacturers based on the shipping means they use. It is then internally standardized for consistent use to drive simplification and avoidance of any improvisation (ad hoc package) cost.

8.2 PACKAGING

Even in the 21st Century, the data-driven era, when everything is continuously optimized, we still find companies that consider packaging as after the fact. When precious resources and time have been spent to design and manufacture a product with all the rigor known to product development, it is still often left to low skill artisans to create and make the boxes meant to protect expensive equipment. Packaging often becomes the weakest link in a rather rigorous process that brings products to life. Packaging must be

---

2 Ibid.

well-designed and meet stringent requirements like the products it protects. Such provisions should include:

- Environmental conditions: temperature variations, humidity, waterproofing, etc.
- Road conditions: vibrations of varying amplitudes and frequencies, drops, etc.
- Loading conditions: load due to other packages' weight on the road and in a stack at a warehouse, drops during loading/unloading, load imbalance due to handling, etc.
- Environmental protection: disposal after use; a recyclable package would not go to landfills.
- Constraints: for any package that will get into a house through a doorway, it is advised to stay below a given dimension, for example, 30." In the United States, the most popular door widths are between 24" to 36."
- Etc.

The function of a package is to "Protect Load." Just like it was presented in Chapter 5, any product requirements definition should include packaging requirements. The requirements of a robust packaging should be defined to protect the load. An intentional design can protect the load and still be cost effective.

A modular product architecture offers the unique opportunity to leverage the clearly defined product structure to think through from the system's design to its modules, components, and even parts.

8.2.1 Inbound Flow Packages

For a company that performs its final assembly in the factory, the inbound flow will carry the systems' items of varying complexity from raw materials to parts delivered by bulk or modules pre-assembled from external sources. No matter the level of pre-assembly of what is delivered to the assembly line on the factory floor for the final system integration, the packages should be designed and contain the Packaging Bill of Materials (PBOM). The list should be unambiguous, and the content of the package verifiable. Such a discipline simplifies the verification of a component or a module through visual inspection. Modules' input and output from the system have already been validated due to their independent nature, defined earlier in Part I of the book. So, all the modules delivered to the assembly floor should have their bill of materials with mistake-proofing (Poka-Yoke in Japanese) instructions for inadvertent errors free, assembly, or installation.

Example

Let us assume XYZ Inc., a city bike manufacturing company that has made the strategic decision to manufacture its bike's Subsystem 1 in house and outsource all the other modules and subsystems to its Subsystem Supply Partners (SSP). The company will only have machines to produce the components of Subsystem 1's modules. There will be no packages for the following modules since they will be produced locally:

- Frame
- Fork
- Front Fender
- Rear Fender

Based on Table 8.1, subsystems 2 to 10 will all come in packages. Each SSP will deliver a subsystem. Some may even deliver more than one. The advantage of a modular environment is that all the packages can use standard numbering. Modules' packages of a subsystem can all be standardized regardless of the module instances and variants.

SS ID	Subsystems	Mod. ID	Modules	Components
1	Frameset	1.1	Frame	Frame, Rear shock absorbers
		1.2	Fork	Fork, Bearings, Nuts, Front shock absorber
		1.3	Fender	Front fender, Rear fender
2	Propulsion	2.1	Pedals	Right Pedal, Left pedal
		2.2	Motion transmitter	Chainrings, Crankarms, Spider, Chain
		2.3	Speed selector	Cassette, Front derailleur, Rear derailleur
		2.4	Motion converter	Crankset
		2.5	Power converter	Drive
3	Wheels	3.1	Front wheel	Rim, Hub, Spokes, Bearings, Tube, Tire
		3.2	Rear wheel	Rim, Hub, Spokes, Bearings, Tube, Tire
4	Bike Control	4.1	Steering assembly	Handlebar, Headset, Stem, Grips
		4.2	Acceleration lever	Acceleration Lever / Shifter
		4.3	Stopper	Stopper
5	Brakes Assembly	5.1	Front Brake	Disc, Caliper, Pad
		5.2	Rear Brake	Disc, Caliper, Pad
		5.3	Brake cable	Front cable, Rear cable
		5.4	Brake levers	Front lever, Rear lever
6	Dashboard	6.1	Speedometer	Speedometer, Wire, Wheel hub connector
		6.2	State of charge indicator	Display, Wire, Battery clamp
7	Load Carrier	7.1	Seat	Seatpost, Saddle clamp
		7.2	Seatpost	Seatpost, Saddle clamp
		7.3	Luggage carrier	Luggage carrier
8	Power Generation	8.1	Generator	Generator, Generator clamp
		8.2	Engine	Engine, Clutch, Exhaust pipe
		8.3	Tank	Container, Filter, Cap, Hose, Tank clamps
		8.4	Fuel	Fuel
		8.5	Electric motor	
9	Energy Storage	9.1	Battery pack	Battery, Battery clamp, Charge electronics
		9.2	Battery charger	Battery charger
		9.3	Charge supervisor (SW)	State of charge software
10	Lights	10.1	Headlight	Headllight, Headlight clamp, Wire
		10.2	Rear Light	Rear Light, Rear light clamp, Wire

Table 8.1 Bicycle Root BOM and components

The Packaging Bill of Materials (PBOM) is derived from the Root BOM. An illustrative example is represented in Table 8.2.

SS ID	Subsystems	Mod. ID	Modules	Components	Packages
1	Frameset	1.1	Frame	Frame	P1-M1.1
				Rear Shock absorber	P 2-M1.1
		1.2	Fork	Fork	P 1-M1.2
				Bearings, Nuts, Front shock absorber	P2-M1.2
		1.3	Fender	Front fender, Rear fender	P1-M1.3
2	Propulsion	2.1	Pedals	Right pedal, Left pedal	P1-M2.1
		2.2	Motion transmission	Chainrings, Crankarms, Spider, Chain	
		2.3	Speed selector	Cassette, Front	
		2.4	Motion conv		

Table 8.2 Bike PBOM (Illustrative example)

Following Table 8.2, some of the packages received in the bike assembly factory could be standardized as follow:

A. Subsystem 3 - Wheels

Since the bikes will be assembled in the factory, there is no need to box the wheels by a pair for a specific contract. Therefore, they can be shipped in boxes with several wheels. The box and its label could be as shown in Fig. 8.3.

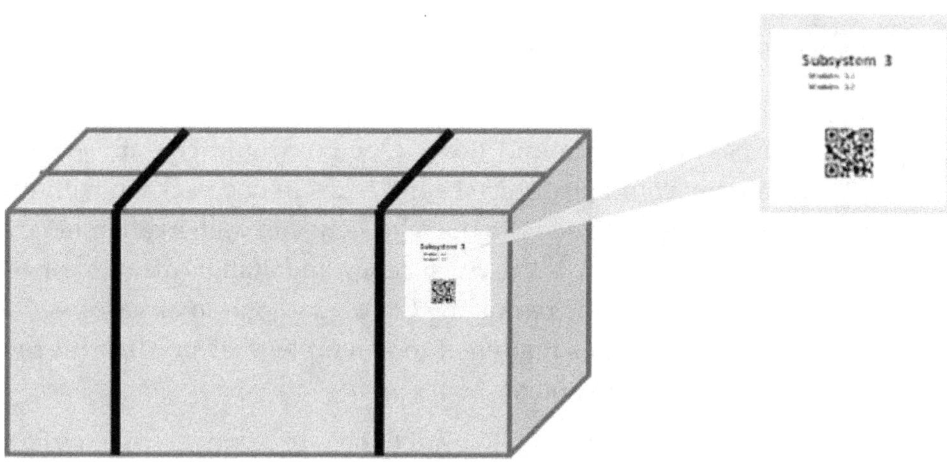

Fig. 8.3 Subsystem 3 – Wheels and Label

B. Subsystem 7 – Load Carrier

If instead, subsystem 7 Load Carrier was considered, it would be shipped with three different packages with each containing several:

- Seats
- Seatposts
- Luggage carriers

Each box would contain the same variants of a module from subsystem 7. The variants of the same module could be by shape or color. The content of each box would be on the label.

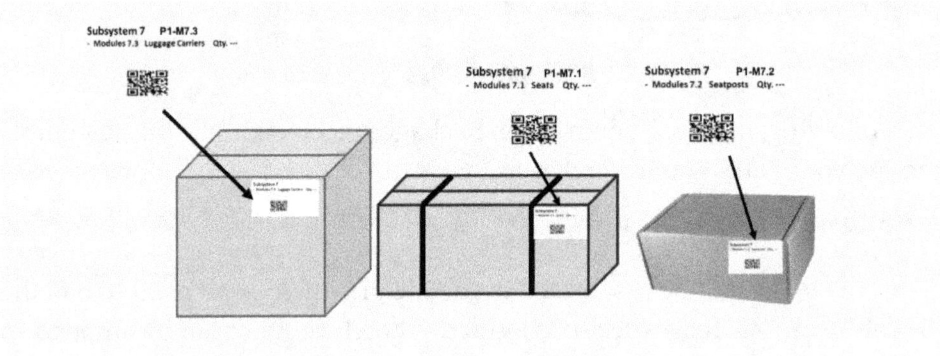

Fig. 8.4 Subsystem 7 with each of the modules in one dedicated box

8.2.2 Outbound Flow Packages

Assuming the same XYZ Inc. company, fully assembled bikes would be shipped in boxes for its outbound flow. One box could contain several bikes. The company could also ship several bikes in one package without the wheels to minimize its size. Then the wheels would be in a different box.

A company's opportunity is the consistency and standardization of all the packages that make up the packaging bill of materials in all cases.

In general, for companies that need to assemble their products at the jobsite, a few differences are noted.

- Inbound flow subsystems need to be delivered by contract from the SSP. Each subsystem boxes will have labels.
- There will be the contract number in addition to the standard subsystem number and the modules contained in the boxes. The same information

will be on the readable tag (bar code, QR code, etc.).

- Upon delivery from the SSP, the inbound flow boxes will be consolidated by contract in the factory's picking area or a warehouse. The warehouse could be a logistics center or a distribution center.
- The list of the consolidated material per contract will be the Shipping Bill of Materials.

In conclusion, each package will have its PBOM, and each contract will have its SBOM.

8.3 TRANSPORTATION

Nearly all the goods produced need transportation. Elements that are transformed or require integration will be part of the inbound flow and transported to the manufacturing or assembly locations. The finished goods will then go to the customers (consumers, distributors, dealers, etc.). Even if the book does not cover all possible produced systems' movements, let's explore some of the modular products' opportunities.

8.3.1 Inbound Transportation

Packages received by manufacturing companies are generally made of raw materials, semi-finished components, finished modules, and even finished complete subsystems, mostly when shipped from an SSP. There is a distinction between temporary and final packaging. Packages delivered by suppliers with contents integrated into a final assembly are opened in the process and considered temporary. Their design must protect their content and be recyclable to avoid landfill in cases for raw materials, semi-finished components, or finished modules that need to be either transformed or integrated into a subsystem or the final system.

Final packages from an SSP are designed and consolidated with other system packages before the outbound transportation. They are loaded as received and not be opened in the process, as is often the case for fully assembled systems at a customer site. The labeling and numbering of such packages should already reflect their subsystems and even modules for simplicity, clarity, and traceability.

8.3.2 Outbound Transportation

While for the inbound flow, companies deal with their suppliers, the outbound flow is more complicated. Fig. 8.5 depicts some of the possible routes a product could take to reach the consumers. In the process, there may even be value-added work performed by some of the channels responsible for delivering to the customer. Packages may change as they go through a third party, due to their content integration, before reaching the consumers.

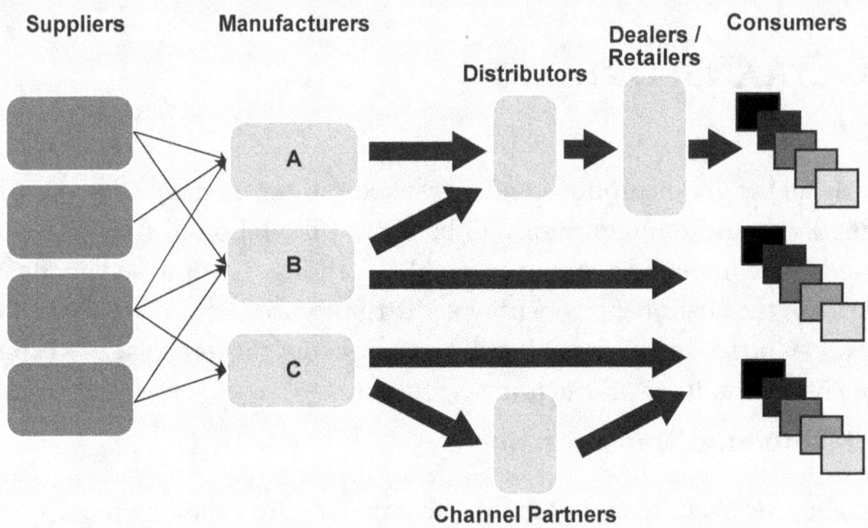

Fig. 8.5 Complexity of Outbound Flow

A couple of distinctions to make are between companies that sell directly to consumers and those that reach them through a third party. Some companies do both. They have their own channels as well as work with partners. For now, however, we will distinguish between direct sales and indirect sales.

For companies that sell directly to consumers, they have the freedom to design their packages and follow the nomenclature they defined for the PBOM and the SBOM. They can also standardize the carriers they need based on the boxes' size, delivery time, and other factors.

If the company sells through a third party, the further distinction that can be made is whether it sells a finished product through partners or a product that still needs value-added work before its delivery. When finished products are delivered, even if it is through a third party, the manufacturer has the freedom to design its packages with their PBOM and SBOM. It also has the freedom to define the carriers from its facilities to the partners. However, transportation means to the consumers fall to the partner and should be agreed upon by the parties. In such a case, the manufacturing companies should still define and recommend how their products should be transported as finished goods to the consumers to maintain their integrity and security.

Suppose a third party performs value-added work before delivery to consumers. In that case, the Value-Added Resellers (VAR) have the ultimate responsibility for what is delivered. Still, the Original Equipment Manufacturers (OEM) should take all the precautions to keep the product's integrity, especially if most of the subsystems come from them.

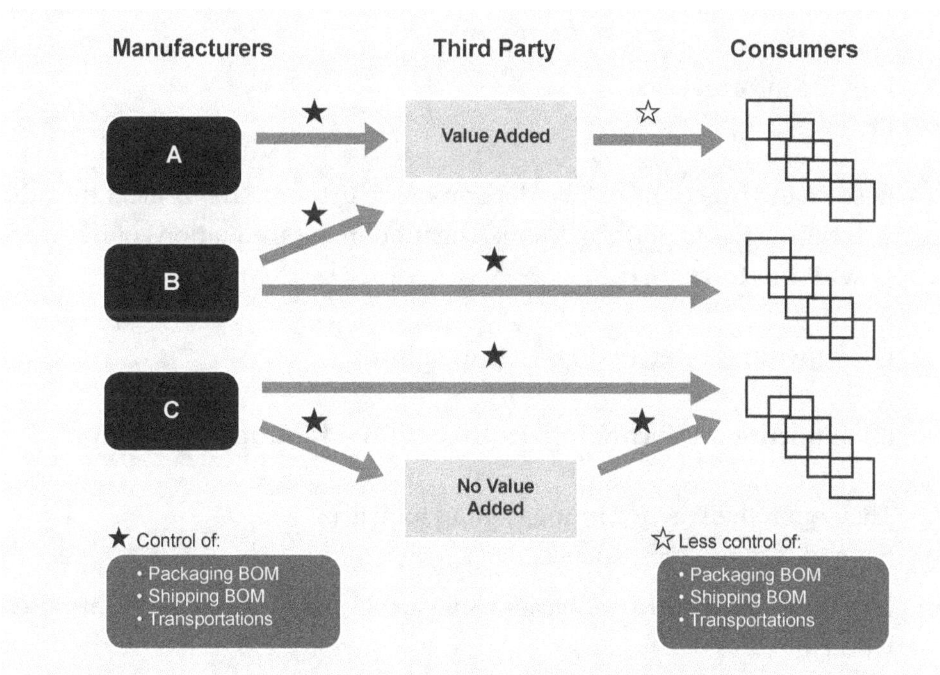

Fig. 8.6 PBOM, SBOM and Carriers Control

In Fig. 8.6,

- Company A, which sells through a value-added third party, does not fully control the PBOM, SBOM, or the carriers from its partner to the consumers.

- Company B has both options. It sells directly to the consumers and has full control of the logistics. But it has less control over what happened to its products after their delivery to a third party.
- Company C has full control in both cases since its third party does not add any value.

8.4 WINNOVEX™ IN LOGISTICS

Any valuable good produced in one place to be delivered in another needs to be protected to preserve its integrity. Among the functions required to achieve such a goal, we could consider:

- Protect load
- Handle load
- Transport load
- Provide information
- Etc.

Each of the functions can be decomposed further to help select the appropriate technologies to perform them. The fundamental equations of innovation apply with full force.

(1) Functions + Technology = Innovation

(2) (Functions + Technology) x Execution = Winning Innovation

They open the way for innovation in logistics.
Several technologies, such as,

- Materials: cardboard, polymers (plastics), metal (steel, aluminum, etc.), wood, etc.
- Bonding: glue, Staples, screws, welding, etc.
- Temperature control: cooling, refrigeration, dehumidification, etc.
- Information transfer: bar code, quick response (QR) code, RFID, etc.

These technologies are only a few among others that can be mapped to the functions with an excellent execution to deliver winning innovations of equipment and information flow management between suppliers and manufacturers and between manufacturers and consumers.

8.5 LOGISTICS SUMMARY

The ultimate judges of a product are the consumers. Therefore, manufacturing companies have to guarantee the integrity of their products until they reach their final owners. After the hard work of design, manufacturing, and assembly of their products, companies should not downplay their logistics. Unfortunately, packaging is often overlooked and is an afterthought. When precious resources and time have been spent to design and manufacture a product with all the rigor known to product development, it is still often left to low skill artisans to create and make the boxes meant to protect expensive equipment. Therefore, packaging often becomes the weakest link in a rather rigorous process that brings products to life.

In the chapter, we have explored some packaging characteristics and how they can be optimized in a modular product environment, both with the inbound flow from the suppliers to the manufacturers and to the consumers in the outbound flow. As the function of a package is to "protect load", any well-designed packaging must meet requirements that include:

- Environmental conditions: temperature variations, humidity, waterproofing, etc.
- Road conditions: vibrations of varying amplitudes and frequencies, drops, etc.
- Loading conditions: load due to other packages' weight on the road and in a stack at the warehouse, drops during loading/unloading, load imbalance due to handling, etc.
- Environmental protection: disposal after use; a recyclable package would not go to landfills.
- Constraints: for any package that will get into a house through a doorway, it is advised to stay below a given dimension, for example, 30." In the United States, the most popular door widths are between 24" to 36."
- Etc.

Beyond the packages, their transportation also deserves special attention. We have explored the cases when the finished goods are delivered directly to the consumers and when they must go through a third party and whether the partners sell direct or add value before selling.

The same attention and rigor given to the design and manufacturing of a product should be extended to their packaging, transportation, and information flow to protect their value until they reach the final customers. The same principles of needs, requirements, functions, and technologies discussed for product design and the diligence used to design and deliver on those, must be applied to ensure the simplest, most reliable logistics structure. The effort should be coordinated by OEM's with their logistics partners.

The fundamental equations of innovation find their application here, as well, and open new opportunities for:

- Product integrity,
- Product differentiation, and
- Customer delight

9

Field Installation

In the value chain, before any product's delivery to the final customer, it either needs to be fully assembled in a factory and delivered as a plug and play or installed at the customer's site. Several old school companies have field mechanics dedicated to the installation of their equipment. These good technicians are often given the installation manual, which they follow step by step to assemble the components at a customer site. The company would then send field systems engineers who have the big picture of the system, inspect, adjust any out of tolerance parts, run all the software applications, make sure the equipment operates per requirement, and complete the commissioning before handing it over to the customer. Such a process has served many industries where the products were mostly electromechanical assemblies. But in the 21st Century, systems are evolving rapidly with a lot of software modules performing functions that were managed through physical entities. Systems with push-buttons and knobs have been replaced by touch screens. We open doors with apps on smartphones with no mechanical contact. This trend calls for companies to review their installation practices and adopt new ways to take advantage of the digital world's opportunities.

Every company defines its assembly or installation process to maintain the product's quality and cost. For medium and large-sized products that need to be transported and fully assembled at a site, the installation time can drive the total cost competitiveness beyond the materials cost. The process warrants a closer look and observation to better understand the underlying factors and influences. Installation cost assumes particular relevance in high labor cost countries such as the United States, Canada, Germany, Japan, etc.

Just as every module that goes into a product is designed to meet rigorous requirements, and the system design includes the manufacturing process, the installation processes should also be part of the consideration. The installation processes should not be defined after the designers have already made all the critical decisions and locked the design. Companies that adopt a concurrent product and process design, as shown in Fig. 9.1, avoid costly mistakes due to miscommunications and missing linkages between competencies. Successful companies create Integrated Product Development (IPD) teams that work together on:

- Module design,
- Module manufacturing process,
- Module installation process,
- The review and revision of the system assembly process (if assembled in the factory),
- The review and revision of the system field installation process,
- The integrity of the whole system.

Such a concurrent effort offers the perfect ground for the verification of the module interfaces. It should be supported by the modeling and simulation of the whole system, which will be reviewed in Chapter 12. With its manufacturing and installation processes, the physical system should be designed and digitally modeled and simulated concurrently.

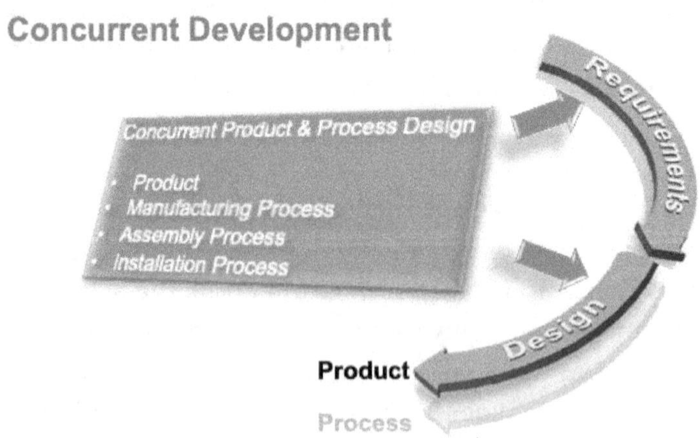

Fig. 9.1 Concurrent Product & Process Development

Manufacturing is known to be rigorous, and its processes are continuously improved to meet both takt time[1] and cycle time.[2] The process cost of each item is known and tracked. A sound installation process should be the same. It should be predictable, based on a rigorous process. To succeed, premier companies have:

- The system architecture with all clearly defined subsystems and modules,
- The system requirements that include the installation process requirements,
- The full digital model of the system to be simulated from end to end,
- The Packaging Bill of Materials (PBOM)

Once all these essential documents and tools are in place, a system's installation after its delivery at a site should be no different from its assembly in a factory-controlled environment.

Following is a brief description of each of the documents and tools mentioned above.

9.1 SYSTEM ARCHITECTURE AND PACKAGING BILL OF MATERIALS

The system's functional architecture defines the Root BOM with all the subsystems and their generic modules. Its importance for installation is in the nomenclature of all that is delivered at the site. Each subsystem and its modules have their identification numbers as defined by the architecture. Such an arrangement allows the packaging bill of materials to identify each package and its content. Before the installation, each box should be placed in a predefined and possibly standardized location on the set. Details of the order in which the packages should be opened should not be left to the installers. They should be part of the instructions that are the result of the

1 Takt time is the term used to describe the required product assembly duration needed to meet demand. In a sequential unit production, it is the average time interval between the start of production of unit #1 and the start of production of unit #2.

2 Cycle time is the total time of a unit of work from the beginning to the end of a process (including process time and delays)

fully modeled and simulated system. The PBOM, as shown in Fig. 8.2, is important and is leveraged to label the packages. The labeling immediately signals if any module or any of its components is missing when the box is opened. Upon delivery, the material is inspected, and each package is carefully positioned as predetermined in the installation phase. Upon observation that a component or module is damaged, its standard part number can be leveraged to order its replacement.

9.2 SYSTEM REQUIREMENTS

The importance of complete and traceable requirements cannot be emphasized enough. They need to cover all the aspects of the system, be thorough and traceable, as mentioned in Appendix C (of Chapter 5). Good system requirements should include the installation requirements of any system's constituent, including the modules and the subsystems. They drive the installation process's definition and should leverage Design for Installation (DFI) rules as part of the product design. The practice would significantly reduce any risk of rework after the whole system is installed. Treating field installations with the same rigor on factory assembly lines would give a competitive edge, especially in high labor cost countries.

9.3 SYSTEM DIGITAL MODEL

Whatever product they manufacture and install, it is important for companies to model it digitally. The introduction of a full system modeling, during the design phase, is an investment that companies hesitate to make. Still, once they overcome their doubts and cross the line, they realize that the benefits are simply outstanding. System models have existed in some companies for decades and have been continuously refined along with all the modeling and simulation improvements. In the 21st Century, with the data analytics revolution still unfolding, full system models have assumed even a more relevant role as their connection to live systems drives the wonder of digital twins covered in Chapter 12, dedicated to modeling and simulation. The installation process should be fully simulated by dedicated teams working closely with field technical personal. Any data collected should be fed back and integrated into the digital models to make the process efficient, cost-effective, and predictable.

9.4 SYSTEM INSTALLATION PROCESS ON THE WINNOVEX™ WAY

Identifying the functions to perform to deliver a consistently fast and safe installed product should be at the front and center of the definition of the field work process. Once the functions representing the tasks are set, several technology categories should be considered:

- **Digital technologies**. As described in the previous section, modeling and simulation of jobsite activities can help determine the operations' right sequence. More importantly, immersive virtual reality, for example, can aid the assessment of the process's safety and ergonomic impact on workers in a hazard-free environment before any dry runs and to the field.
- **Field tools**. Installation aids and tools should be designed to augment workers' ability to handle loads and accelerate assembly operations. A technology landscaping should be conducted and the right one selected to meet the installation requirements.

There is ground for innovation, and the fundamental equations (1) and (2) also apply to make a difference.

All the information gathered from the system architecture, the product requirements, and the digital models provide critical input to the definition of a good installation process. As shown in Fig. 9.2, installation can still be created without the right prerequisites, but it would neither be optimized nor optimizable over time. The process can also miss crucial innovation opportunities.

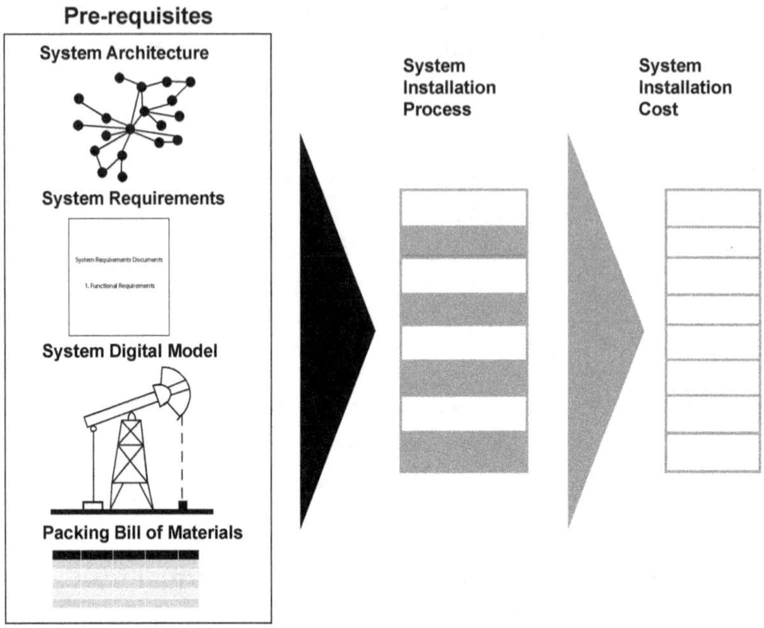

Fig. 9.2 Installation Process

9.5 SYSTEM INSTALLATION COST

Just like in the factory, a good installation process defines the time needed to install each module. The exercise is simulated and executed through practice; then, it is continuously refined based on the field's feedback. Once the process matures and meets its execution requirements, it can be applied in different locations, and the results can be compared for continuous improvement. It is crucial to break-down the installation time to module level to become part of any redesign requirements.

It is not rare to observe design teams celebrate a cost reduction project, only to get negative feedback from the field engineers and product management because their work did not reduce the total install cost, or even worse, it caused it to go significantly up. By extrapolating Table 8.2 and generalizing it to any system, Table 9.1 has been created and has been successfully used by different companies.

Table 9.1 shows a system's total installed cost through its subsystems and modules. As the system's building blocks, the modules' material cost (factory cost) is first determined, followed by their installation times. The total installation time was then converted into the installation cost based on the standard cost of one installation minute.

SS. ID	Subsystems	Mod. ID	Modules	Components	Module Material Cost	Module Installation Time	Module Installation Labor Cost	Total Module Cost	Total Subsystem Cost	Comments
1	Subsystem 1	1.1								
		1.2								
		1.3								
2	Subsystem 2	2.1								
		2.2								
		2.3								
		2.4								
		2.5								
3	Subsystem 3	3.1								
		3.2								
4	Subsystem 4	1.1								
		1.2								
		1.3								
5	Subsystem 5	5.1								
		5.2								
		5.3								
		5.4								
6	Subsystem 6	6.1								
		6.2								
7	Subsystem 7	7.1								
		7.2								
		7.3								
8	Subsystem 8	8.1								
		8.2								
		8.3								
		8.4								
		8.5								
9	Subsystem 9	9.1								
		9.2								
10	Subsystem 10	10.1								
		10.2								
			Total System Cost						Sum (Rows 1-10)	

Table 9.1 Installation Time and Cost

It is important to point out that some companies include the logistics (packaging, loading, shipping, and transportation) cost to estimate the installation minutes cost while other companies keep them separate and add them when computing the total installed cost.

9.6 FIELD INSTALLATION SUMMARY

The rigorous processes followed during product design and manufacturing could be easily compromised if the installation process is not defined and executed at the same level of importance. Product design must be done concurrently with manufacturing, supply chain, and installation processes, as shown in Fig. 9.1.

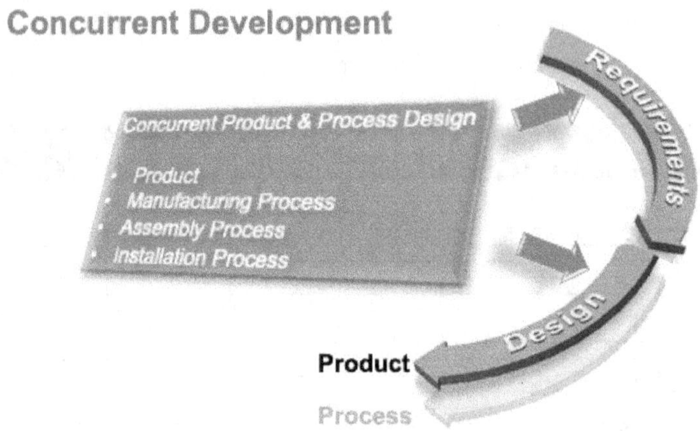

Fig. 9.1 Concurrent Product & Process Development

In premier companies, a sound installation process design is supported by:

- The system architecture with all clearly defined subsystems and modules,
- The system requirements that include the installation process requirements,
- The full digital model of the system to be simulated from end to end,
- The Packaging Bill of Materials (PBOM)

Those essential documents and tools lead to an efficient, cost-effective, predictable, and standard installation process where the total installed cost of each module, subsystem, and, subsequently, the system is defined. Such a standard method for the company can be replicated in all the locations where it operates.

Defining the field workers' functions and mapping them to selected technologies can augment their safety and abilities to deliver world-class performance keeping the company on the leading edge.

Companies need to be intentional about installation processes and treat them with the same rigor found on factory assembly lines. Doing so will give them a competitive edge, especially in high labor cost countries.

10

Aftermarket and Service

Any high-value asset needs to be kept operating ideally for a long time at its initial performance, close to it, or at a minimum acceptable level, to get a good return on the investment during its lifetime. Such a performance requires routine maintenance and timely repair of any malfunction or damage to any system module or component. Contracted service after purchasing equipment from a manufacturing company has evolved over time with the sophistication and efficiencies of the continuous infusion of technologies.

Services fall under two broad categories. The first category is the one offered by manufacturers as aftermarket and equipment service, and the latter is about general service business companies. Such a distinction will allow the review of how *The WinnoveX™ Way* applies to both types of services.

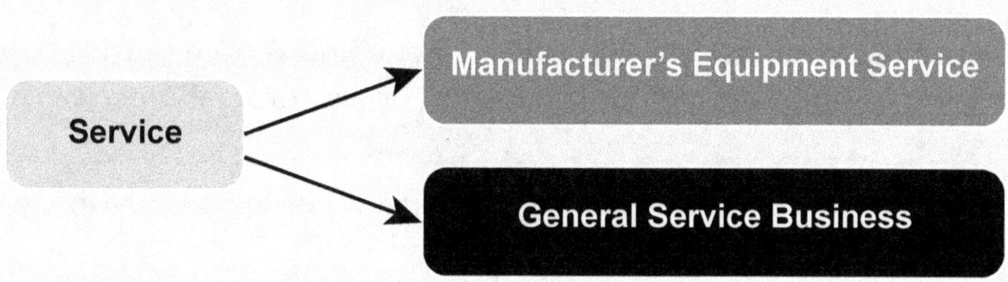

Fig. 10.1 Distinction of Manufacturers' Equipment Service from General Service Business

10.1 MANUFACTURERS' EQUIPMENT SERVICE

Several studies were conducted on equipment service. Some of them went all the way to the second half of the 19th Century to track manufacturing companies. They made some interesting discoveries, in particular, the research conducted by Professor Roger W. Schmenner of Kelly School of Business (Indiana University) found that "the bundling of manufactured goods to downstream-available services was led by companies with new products but with no great manufacturing strength, as a way to establish barriers to entry."[1] Even if it was, initially, triggered by the threat of new entrants, aftermarket service is now an important way for customers to protect their assets. There are many ways by which aging equipment can face the degradation of its performance. The source could be mechanical, electrical, or even software related. Mechanical failure could be related to a multitude of causes. A few examples could be:

- Fatigue of moving components,
- Wear of ill lubricated rotating components,
- Heat generated by insufficient equipment ventilation or cooling,
- Corrosion of critical parts, etc.

From an electrical failure standpoint, we could observe:

- Defective insulation,
- Loose connections,
- Power surge from a regular line or lightning,
- Electrostatic discharge or ESD and improper grounding, etc.

Functions performed by software are not immune from causing failure of an equipment. Just like for mechanical or electrical failures, we could observe:

- Software not sufficiently debugged due to implementation rush,
- Inadequate updates,
- Operator misuse due to insufficient training,

1 Publication *"Manufacturing, Service, and their Integration: Some History and Theory,"*Prof. Roger W. Schmenner Kelley School of Business Indiana University, August 2008, p.1

- Fault in an electronic component, etc.

Any of those examples could lead to a system failure and cause performance degradation and even downtime with serious financial consequences. To avoid unplanned stoppages or equipment inefficiencies, original equipment manufacturers who provide the aftermarket services, and those that have developed the competencies to service someone else's products are actively involved in maintenance and repair.

10.1.1 Lowest Replaceable Units

The notion to introduce before exploring repair and maintenance is one of the Lowest Replaceable Unit or LRU. The process of fixing damaged or faulty equipment often requires the replacement of some items. Once the root cause is determined, a surgical type of effort should be executed to determine what needs to be fixed or replaced to avoid costly interventions. For years, product development experts have included in the product requirements the definition of the items to be replaced at the lowest level without compromising the system's safety, functionality nor performance; Such items are known as the Lowest Replaceable Units or LRU's. They cover a wide range of items at various levels. In a modular environment, an LRU could range from a part, a component, or even a whole module. They are predefined to allow their quick replacement. They are designed as naturally part of the system, simulated, built, and tested to assure the system's integrity with their use.

10.1.2 Repair

The Merriam-Webster dictionary defines the verb "to repair" as "to mend, meaning to put into good order something that is injured, damaged, or defective. This implies making whole or sound something broken." Human-made entities need to be kept healthy with interventions to make them whole again after a fault or damage like any living being. Equipment repair is a business segment on its own as part of the vast aftermarket opportunities.

10.1.3 Maintenance

Unlike a repair that requires an intervention to correct a fault during the use of a product, maintenance is more intentional. There are several types due to the nature of the activities involved. For example, we could mention:

- **Periodic maintenance**: known as Time Based Maintenance or TBM, it is one of the pillars of Total Productive Maintenance (TPM) and is usually performed by the equipment operators. The activities could include visual inspection, cleaning, lubrication (oil and grease), removing debris (in case of machining centers) or dust, etc.
- **Preventive maintenance**: to prevent an unexpected failure, periodically planned and engaged dedicated people inspect the equipment. They usually receive more training than the operators. Their work involves replacing any parts that have been assessed for potential failure before the next inspection.
- **Predictive maintenance**: moving beyond the periodic inspection, this important type of intervention is due to a predicted failure resulting from either modeling and simulation or a data-driven analysis of some critical modules' condition through established and monitored parameters. By tracking selected variables such as temperature, sound pressure level, vibration, power consumption, or others, for established "normal" values, algorithms are set to predict the next scenarios of an equipment's performance or even time to failure.

By performing any of these maintenance categories on a physical system, the goal is to keep it up and running with no unwanted interruption and at or close to the expected peak performance. Predictive maintenance is the most advanced of the type cited.

Companies with a high conversion rate from new equipment sales to service contracts would find it worth investing in preparing their products for predictive maintenance. The impact on Total Cost of Ownership (TCO) for their customers could be a differentiator in a competitive market.

Such investment and effort should start from the system's design phase. As it was emphasized in various chapters of the book,

- System requirements need to be well captured and allocated to the subsystems and their modules.
- Module interfaces need to be defined and, at least, those of the core and hub modules need to be captured in ICD's,
- Good digital models of the system need to be in place,
- Both system requirements and models need to be validated through internal testing and field feedback data analytics

Best in class companies make those investments and get significant savings along with the product's life from their service business.

Fig. 10.2 shows predictive maintenance on a modular product (b) with an extended useful life and performance close to its peak for many more years compared to a non-modular system (a).

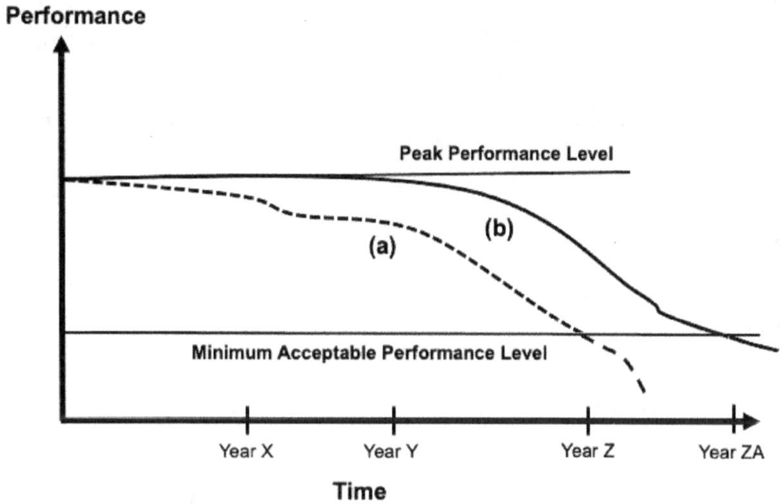

Fig. 10.2 Performance over time due to maintenance choices

In Fig. 10.2, which is conceptual, Graph (a) represents a non-modular system that is well maintained. But, despite a good maintenance schedule, it still shows performance degradation over time. Beyond its components' normal aging, the system struggles to adopt new technologies after Year X without intentionally designed interfaces. After Year Y, it is only left to repairs. It continues to degrade until its performance reaches the minimum acceptable level before its disposal at Year Z. Along their life, non-modular systems are maintained with the same technologies that generated them. Any attempt to upgrade them with new technologies could involve replacing many modules due to their dependencies on one another. So, the upgrade cost would exceed the benefits in life extension and performance. For example, golf carts produced with lead-acid batteries would not get a competitive cost vs. benefit with their conversion to lithium-ion batteries. It would simply be better to buy a new vehicle unless the right interfaces were already designed in place.

For Graph (b), the system is assumed to be modular and has some fully kept up to date digital models. Through a combination of physics-based and data-driven models, predictions would trigger the interventions on the system. Its modular architecture would benefit from module upgrades in technology through intentionally defined interfaces. It would continue to perform at or close to its peak performance for more years than the one represented by graph (a).

Let us consider the *WinnoveX*™ equation: **(F+T) x E = WI**, presented in Chapter 6,

Where,

F= Function,
T= Technology,
E= Execution and
WI= Winning Innovation,

The OEM's have the opportunity to inject, in the course of the product's life, multiple generations of technologies to perform the functions driven by the customers' needs, as it was mentioned earlier in the book. With the golf cart conversion example described earlier, having a modular design, the established architecture would pinpoint the modules to replace with clear interfaces in an eventual battery conversion without the burden to redesign most of the vehicle.

In general, predictive maintenance would be aided by the system's architecture through its modules since the Lowest Replaceable Units (LRU's) would have been defined with the modules at the time of the product design and validation, with the interfaces. Also, having a good architecture in place, the serviceability of any complex product can be simulated. In particular, the compliance to the following category of requirements can be modeled and simulated:

- LRU removal and installation without disassembling large sections of the system,

- Ergonomic requirements for worker safety (since at the service site, not all the factory tools and aids would be present),
- The predetermined functional performance of the modules and their LRU's to guarantee the system's integrity after its installation,
- The system's upgrade with next-generation modules based on new technologies

To keep any system healthy and operating as close as possible to its peak performance in a technologically dynamic world, determining and proactively managing its functions are more critical than ever. As mentioned, several times in the book, a system's functions evolve at a much slower pace than the technologies that perform them. Therefore, the system's functional architecture is one of the most precious documents of any company and continues to be relevant even after the product's release. The metrics of the functional requirements defined from the architecture should be kept up to date with the technologies leveraged to perform them. Through modeling and simulations, life predictions can be made on a system's modules to drive future upgrades by leveraging technology advancements.

A couple of centuries-old practice that is still in force today and applicable to non-modular systems' repair and maintenance is the replacement of an old faulty part with another one originated from the same design. In the 21st century, we should be replacing a damaged or faulty part with one that performs the same functions and with better technology. Some OEM's continue to produce their service parts with the same technologies that originated them. Such a choice is understandable and makes sense in existing systems due to their poorly designed and managed interfaces. Still, it is hardly conceivable that new systems continue to be designed, even by large companies, to be non-modular with all the cost carrying implications during the total product life cycle. Any company with a strong desire and goal to be relevant and competitive in the 21st Century must embrace modularity, not only for its new equipment but also for its service business.

10.1.4 Equipment Service in the 21st Century

A lot has been and continues to be said about industry's big revolutions, 1.0, 2.0, 3.0, and now 4.0. As mentioned in Chapter 1, the convergence of some new technologies is offering more than ever, the opportunity for companies to benefit from mass customization with significant cost savings for a healthy bottom line and top-line growth. Even though this topic will be explored in Chapter 14, a few observations can be made on equipment service during industry's fourth revolution. In this cyber-physical world that is unfolding at warp speed pace every day, equipment connectivity, or, in general, systems' connectivity is vital. Through their control of subsystems and health monitoring sensors, systems provide data in amounts expressed in numbers that so many zeros behind a digit can only attempt to convey. Data collection alone may not move the needle, but when the right algorithms are put in place, and the data reveal insights, systems assume a different meaning. Here, the leading-edge companies are increasingly creating their systems' digital twins by taking their modeling and simulation capabilities to a new level. Their systems' health is monitored in real-time through the analyses of data collected at manageable intervals. Again, modular systems offer an edge over other design types. Modules designed as independent entities are now being conceived with additional self-health check capabilities. So, each critical module's health can be monitored real-time, and the system can flag any anomaly before it happens. More importantly, an anomaly such as a vibration can quickly be tracked to the source through the vibration signature of all the parts, components, and, subsequently, modules involved. Using Artificial Intelligence (AI), the systems are also evolving toward perceiving their conditions and making the appropriate decisions autonomously with little to no human intervention. The future is opening new opportunities beyond predictive maintenance into what could be defined as cognitive maintenance. Once the technologies that define it are available, they will bring the companies willing to invest in leading-edge technologies to deliver "evergreen" systems with the potential of an endless series of metamorphic upgrades.

10.2 GENERAL SERVICE BUSINESS

Companies that do not manufacture products but provide services to other businesses or consumers define the general service business. The question to explore in this section is how *The WinnoveX™ Way* applies to them. Even though they provide intangible products, they perform important functions that determine their values to their customers. Also, they leverage technologies that drive innovation when mapped to their functions. Therefore, the

WinnoveX™ equations apply to them as well. We will explore one case to illustrate how the application of the equations goes broadly beyond physical systems.

10.2.1 The Postal Service System

When considering the Postal Service, one can say that it has two Operational Mission Functions or Main Purpose Functions:

- Connect People, and
- Deliver Parcels

Before diving into the technical implications of those two functions, it would be fun to share a few fantastic historical facts. History books reveal that postal systems were created by the necessity to send messages to people living in a different place that the sender could not easily reach. When writing was invented, a written message could be carried from one person to another. Discoveries in Egypt have shown that centuries BC, the Pharaohs were already using couriers to send messages to their representatives throughout their land. They were using papyri. All the civilizations with a writing ability have shown different ways of communicating within their empires. The system has evolved with the protection of the written messages, giving birth to envelopes made of different sorts of materials with the function to "protect the message" (which could be decomposed to protect against physical damage and prevent tampering). If we fast forward to explore what could be called the "modern postal system", we will mention the establishment of the postal system in Paris (France) by Jean-Jacques Renouard de Villayer. In 1653, he set up mailboxes with the function to deliver messages placed in his envelopes. In 1837 England, Rowland Hill furthered the postal service by introducing the adhesive stamp, enabling the first stamp postage system to be established in 1840. Toward the end of the 19th Century, the Universal Postal Union was established. Its creation allowed the delivery of messages from one country to another. The United States established the United States Postal Service (known as USPS) in 1775 and appointed Benjamin Franklin as the first postmaster general. Today we have some private package delivery services like FedEx, United Parcel Service (UPS), DHL, and Amazon.

The functions and technologies involved in the postal service are summarized in Table 10.1, drawn to give a sense of some of them. They are not a complete list but a view that will allow a few observations.

Mission / Purpose Function	First Level Functions	Second Level Functions	Technologies Involved
Connect people	Deliver mail	• Collect mail • Transport mail • Distribute mail • Etc.	• Paper technologies • Printing technologies • Sorting technologies • Transportation technologies • Telegraph technologies • Fax technologies • Email technologies • Texting technologies • Social media technologies • Etc.
	Connect voice	• Receive call • Transmit call • Establish conversation • Etc.	• Telephone technologies • Switchboard technologies • Voicemail technologies • Video connect technologies • Social media technologies • Etc
Handle parcels	Collect parcels	• Receive parcels • Condition parcels • Label parcels • Sort parcels	• Packaging technologies • Labeling technologies • Sorting technologies • Transportation technologies
	Transport parcels		• Drone technologies (emerging)
	Deliver parcels	• Load parcels • Release parcels	• Location technologies • Etc.

Table 10.1 Postal Service System Functions and Technologies

- The operational mission functions are decomposed and summarized, as described in Chapter 2, Section 2.2.2.
- The technologies involved are mapped to the second level functions even though they have been kept at a remarkably high level.
- Some of the technologies are no longer in use. They are obsolete even though the functions they performed are still active and continue to express current needs.

The topic of obsolescence deserves some comments. A technology becomes obsolete when new technologies perform the same functions more efficiently at a competitive cost. So, the old technology is discontinued. It is a well-known phenomenon in the electronic world.

It was mentioned several times in the book that a system's functions evolve at a slower pace than the technologies used to perform them. For the postal service, we can note that:

- Emails have totally replaced telegraph, almost eliminated telefax, and partially wounded paper mail.
- The senders now use stamps and labeling from the comfort of their homes.
- Nobody places a call through the post office anymore.
- Parcels management is no longer the monopoly of any state postal service. Private carriers are doing an excellent job and are formidable competitors.
- Drone technologies that are now emerging could provide the last blow that cripples the current way the postal service, most of us grew up with, operates.

Revisiting the *WinnoveX*™ Equation, we would note that in:

(F+T) x E = WI

While the Functions (F) being performed for centuries are still relevant and active, and no significant questions can be raised about their Execution (E), the same cannot be said about the Technologies (T). Some state postal services in many countries have not adapted their technologies to the world around them. Some of them have been hit hard by new ways of performing the functions that they once mastered.

I could go ahead and apply the Postal Service System case to several other service companies and any company or industry.

Technology is a significant variable to consider and investing in it should be a top priority to keep a company competitive and guarantee its growth and longevity.

10.3 AFTERMARKET AND SERVICE SUMMARY

Any high-value asset needs to be kept operating ideally for a long time at its initial performance or close to it, or at a minimum acceptable level, to get a good return on the investment during its product life cycle. The chapter distinguishes between manufacturers' equipment service and a general service company.

Different service types were considered for physical goods, each representing an option available to consumers from manufacturers. The emphasis was put on predictive maintenance since it offers more opportunities for differentiation and growth and applies well with *The WinnoveX™ Way*.

Modular systems offer an edge over other design types. Modules designed to be independent entities are now being conceived with additional self-health check capabilities. Each critical module's health can be monitored. The system can flag potential failure with sufficient notice to effectively address the problem before it happens and before the equipment operation suffers. More importantly, an anomaly such as vibration can quickly be tracked to its root cause through the vibration signature of all the parts, components, and, subsequently, modules involved. Using Artificial Intelligence (AI), the systems are also evolving toward the perception of their conditions and making the appropriate decisions autonomously with little to no human intervention. The future is opening new opportunities beyond predictive maintenance into what could be defined as cognitive maintenance. Once the technologies that define it are available, they will bring the companies willing to invest in leading-edge technologies to deliver "evergreen" systems with the potential of an endless series of metamorphic upgrades.

General service was about the family of companies that do not manufacture products but provide services to other businesses or consumers. The question of how *The WinnoveX™ Way* applies to them was explored. In particular, the Postal Service System case was studied, considering the fundamentals, from its functions to the technologies involved. It could be observed that some states' national postal service systems around the world did not invest in technologies to reinvent their business to stay competitive. Now, they face an upward battle against private companies that are more up to date with leading-edge technologies.

All companies that provide service to consumers must invest in technologies to stay relevant and adaptive to their customers' needs as their business environment or market continues to evolve.

11

Product Management

Most companies that develop and release products have the organizational function of product management. In general, product management provides guidance to the strategy, planning, business case, market requirements, development, release, and marketing of new products based on its knowledge of the company's customers and the competitive landscape. It is a crucial function for setting "What", "Where," and "When" for the product development technical teams who can then focus on the "How." This chapter focuses on the product management function, market analysis based on customer needs, and the company's strengths, weaknesses, opportunities, and threats (SWOT) concerning its competitors. Fig. 11.1 expresses the frame in which product management exercises its responsibilities.

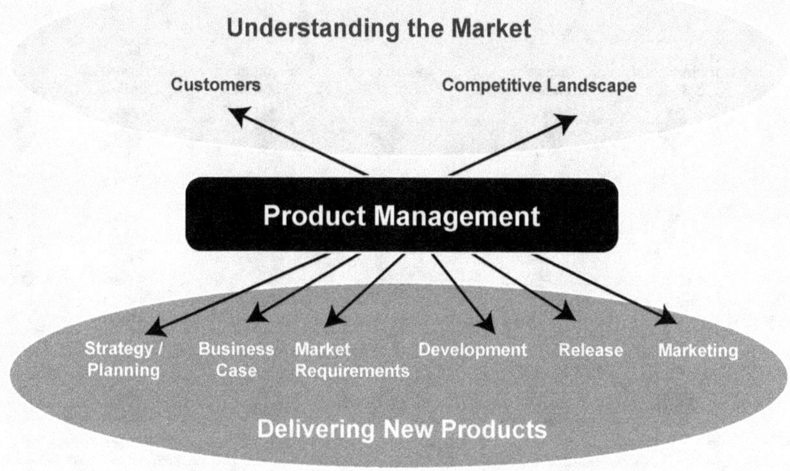

Fig. 11.1 Product Management Functions

Based on Fig. 11.1, we will explore how each item applies to The WinnoveX™ Way elements.

11.1 UNDERSTANDING THE MARKET

An industry is defined by the needs it satisfies, or in other words, the functions it performs since functions express the needs. When discussing the market, it is essentially about understanding the functions necessary to meet the customers' needs. It is also about understanding "how well?" the competitors perform the functions that characterize the market offerings. As expressed in Chapter 6, it is crucial for those involved in product management to understand the different types of unmet needs and where the competition stands in meeting them. Such insight helps to better guide all the other functions involved in product development. Fig. 11.1 shows that understanding the market is to understand both the customers and the competitive landscape.

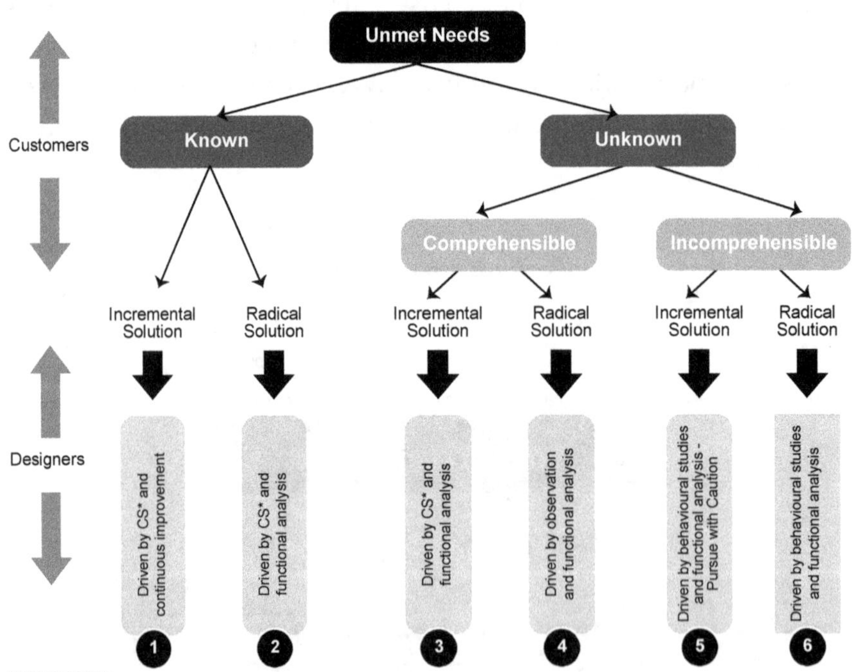

Fig. 6.4 Unmet needs categories and solution types with their drivers

11.1.1 Understanding the customers

A glance at Chapter 6 would help before going more in-depth about the customers. Fig. 6.4 represented again below for convenience, explained that from the customers' viewpoint, the unmet needs could be classified into two big categories:

- The known unmet needs and
- The unknown, unmet needs

Reading further from the figure recalled above, it was interesting to note that when an unmet need is unknown to the customers, it may or may not be comprehensible to them when expressed. When evaluating the solutions, six categories were presented. A customer survey could address some of them, particularly the ones where the need was either known or unknown but comprehensible when expressed to the customers and where all cases were addressable with incremental solutions. Chapter 6 described how to tackle each case. With product management, we should go deeper to the functional level to explore any possible opportunities. Considering the product or portfolio of interest, the functional architecture, made of the modules and their relationships, becomes the reference document that captures the functions performed and the relationship between the modules performing them. Therefore, each module should be reviewed through its functions before any formal customer survey could occur for specific product development.

Here are a few key questions the product managers, the systems architects, and systems engineers should answer before starting any development project:

- What known unmet needs are we currently addressing?
- Through which subsystems and modules are we addressing them?
- What functions are we performing through each of the identified modules?
- What are other functions that could be performed that are not performed today?
- What functions are other industries performing that are transferable to our industry?
- What are some unmet needs that are not yet known to the current customers?

- Would the customers understand them when expressed to them?
- What functions represent them?

The answers to all these questions will help address the categories of unmet needs expressed in Chapter 6 and summarized in Fig. 6.4.

Under the leadership of product management, a team, including systems architects and systems engineers, will conduct a thorough functional analysis to categorize the type of development project the company should undertake. At the end of the exercise, the team will update Table 11.1 (if it already exists) based on the functional architecture of the products or portfolio of products under review or fill in the table for the first time. The result will define the functional library from which the functional requirements should be derived.

<table>
<tr><th></th><th>Subsystem</th><th>Modules</th><th>Known Primary Functions</th><th>Known Secondary Functions</th><th>Imported Functions</th><th>New Functions</th><th>Comments</th></tr>
<tr><td rowspan="9">System or Portfolio</td><td rowspan="3">Subsystem 1</td><td>Module 1.1</td><td></td><td></td><td></td><td></td><td></td></tr>
<tr><td>Module 1.2</td><td></td><td></td><td></td><td></td><td></td></tr>
<tr><td>Module 1.3</td><td></td><td></td><td></td><td></td><td></td></tr>
<tr><td rowspan="3">Subsystem 2</td><td>Module 2.1</td><td></td><td></td><td></td><td></td><td></td></tr>
<tr><td>Module 2.2</td><td></td><td></td><td></td><td></td><td></td></tr>
<tr><td>Module 2.3</td><td></td><td></td><td></td><td></td><td></td></tr>
<tr><td rowspan="3">Subsystem 3</td><td>Module 3.1</td><td></td><td></td><td></td><td></td><td></td></tr>
<tr><td>Module 3.2</td><td></td><td></td><td></td><td></td><td></td></tr>
<tr><td>Module 3.3</td><td></td><td></td><td></td><td></td><td></td></tr>
</table>

Table 11.1 Functions library review summary template

Once the table is ready, the team should take the following actions:

- Set the requirements by leveraging the functions to be addressed and the metrics that need to be met based on the competitive landscape analysis (See next section).
- Review the functional architecture like in Fig. 11.2 to highlight the subsystems and modules impacted or required to create. The result will determine the project's complexity based on the number and type of new modules or modules to update. Remember, more hub modules mean higher complexity.
- Review the instances and variants that need to be updated or added

based on the market size. Understanding the functions will help determine the variants to be re-used, modified, or created new. The magnitude of the project should become clear following the analysis based on re-use, modify, and new design. This information is also useful for the assessment of growth opportunities.

- Leverage modeling and simulation to kick-off the investment analysis that will lead to the business case. No detailed design work is necessary at this stage. Instead, digital models would help clarify the market requirements defined by product management and the design team's potential response to the metrics they need to meet to be competitive.

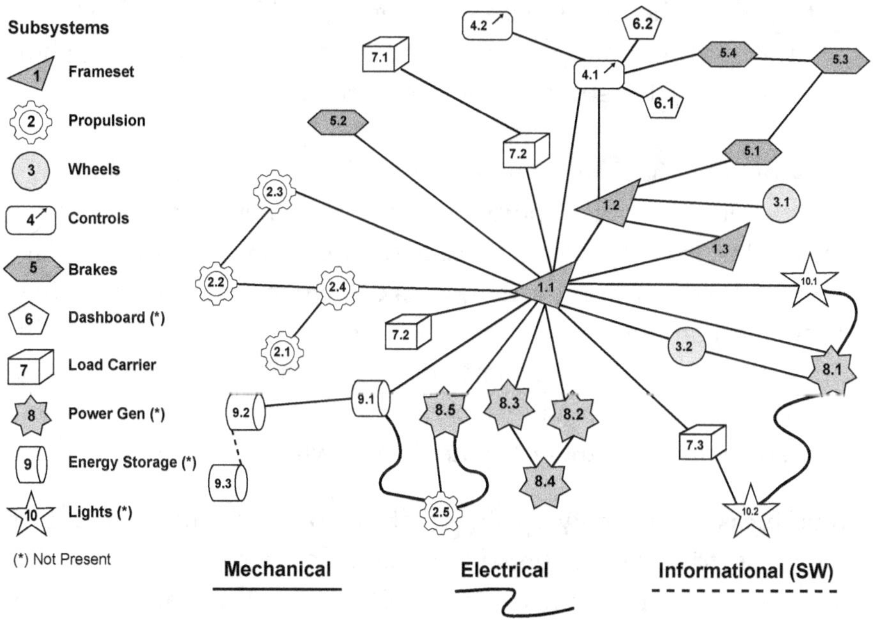

Fig. 11.2 Representative example of a functional architecture (Functional Architecture of a portfolio of bikes Fig. 3.12)

11.1.2 Understanding the Competitive Landscape

Alongside the analysis being conducted by the team led by product management to understand the customers, other products on the market should be analyzed. Unfortunately, most companies make a couple of fundamental mistakes during other manufacturers' products assessment.

1. Competitors' sales documents analysis: it is often observed that some companies limit other manufacturers' benchmarking to collecting their sales brochures, videos, and other officially released material to compile the features offered and then transform them into an internal requirements document. It may seem obvious, but such a practice misses the fundamental reasons that led a manufacturer to offer a feature. The issues are then compounded when documents from several companies are merged to create one that will drive the development of one company's products. Such an approach creates a document based on a mixture of strategies. The "Me Too product" or "They Have It! So, Do We" product strategy hurts more than it helps. The underlying convictions or reasons driving the strategic choices of the companies benchmarked are missed.
2. Tear-downs: they are another method used to understand what is being offered by other companies and how they addressed some critical market and technological issues. It consists of acquiring another company's product on the market as customer and deconstructing it to analyze its components. The practice offers some important insight into different ways of solving some common industry problems. However, the mistake comes from the analysis executed upside down by starting from analyzing the other company's hardware.

In both cases, the analyses should start from the functions performed. The functions drive the functional requirements. If the equipment acquired is new or is still working, the company conducting the analysis should run and test it against its own requirements. Not knowing the competitor's requirements, the analysts should use their own requirement as a reference. They should then capture the metrics expressing "How Well?" the functions are being performed compared to their offerings before the tear-down is conducted. Tear-downs are a vital tool to leapfrog others, but the first step of the analysis should be functional. It should start from the functions before getting to understand how the hardware performs them.

Table 11.2 is a template that can be used to perform a functional analysis of other products offered in the same market. In this case, the company

Table 11.1 Functions library review summary template

conducting the study could simply break the acquired system down along its product architecture.

		Host Company					Benchmarked Comany "X'					
	Subsystems	Modules	Key Components	Known Primary Functions	Known Secondary Functions	Metrics	Module Equivalents	Key Components	Known Functions	New Functions	Compliance to Host Metrics	Comments
System or Portfolio	Subsystem 1	Module 1.1										
		Module 1.2										
		Module 1.3										
	Subsystem 2	Module 2.1										
		Module 2.2										
		Module 2.3										
	Subsystem 3	Module 3.1										
		Module 3.2										
		Module 3.3										

Table 11.2 System Functional Benchmarking Template

The template, which is limited here to only a few subsystems but could include more, allows to gather insight into:

- The functions and some information about the functional architecture,
- The physical architecture, mainly hardware, that allows the assessment of the "module equivalents." Since other products certainly have a different architecture, the group of components that drive a set of functions can be considered a module equivalent in the host's architecture.
- Compliance to the host's metrics. The report would be derived from the full system performance or important parameters measured during its operation. The functions that are better performed can further drive a detailed study.

The insight gathered can help understand some of the features offered in other products to better formulate new development requirements. Systems architects and engineers will also better appreciate the choice of technologies made by the benchmarked company to perform specific functions. Any new function would be identified. In summary, the battle should start from analyzing the functions, not hardware or sales documents.

11.2 DELIVERING NEW PRODUCTS

The development and delivery of new products is a team effort led by product management to fulfill the company's product strategy. There should be a distinction between existing product updates and new product development.

- Existing product updates: existing products should continuously be reviewed and kept up to date. As the customers' unmet needs are uncovered, the existing products are improved to better respond to them. Whether executed to correct manufacturing defects, drive cost reductions, improve the installation process, or simply introduce new features, the continuous improvement projects have one common denominator: they are all done to introduce better ways to perform existing functions through the injection of different technologies. They often provide incremental solutions to known unmet needs or unknown, unmet needs comprehensible to the customers. They belong to categories (1) and (3) from Fig. 6.4.
- New products development: in response to market studies, a company can decide to introduce new products to satisfy:
    - Known unmet needs with radical solutions; category (2)
    - Unknown, unmet needs comprehensible to the customers with radical solutions; category (4) and,
    - Unknown, unmet needs incomprehensible to the customers with radical solutions; category (6).

It is important to note that, as mentioned in Chapter 6, unknown, unmet needs that are incomprehensible to the customers should be very cautiously approached or not addressed at all, with incremental solutions. The incremental gain may not be worth the effort. Solutions to category 6 unmet needs should be thoroughly assessed.

During the investment budgeting, customers' education needs to be accounted for since it would take an effort to have them fully on board. New product developments should be executed in line with the company's long-term strategic product plan and the type of needs to be addressed.

11.2.1 Long Term Product Planning

While companies are busy with executing tactical decisions on day-by-day activities, they should make room for long-term strategic product portfolio planning. The insight gained from the unmet needs of a market segment, when coupled with the megatrends that could impact the industry, allow any company to align its product portfolio with its long-term goals, including financial and market share growths, etc. One of the best ways to analyze a company's long-term product plan is creating product and technology roadmaps. Companies that invest in a road mapping effort are, in general, better positioned than their peers. A lot has been said and written about product and technology roadmaps in literature. However, it is essential to note that their practical use is not as widespread as it should be. Yet, they offer a great framework that enables various functions or departments of a company to visualize their different objectives. Consequently, current and future activities are interdependent in support of the overall company goals. They can better communicate and coordinate their actions toward a set of company performance metrics regarding product development over time. They also help the planning of investments in resources and capabilities in line with the future's vision. A roadmap captures many ingredients, but it should be carefully chosen to avoid a heavily crowded document that loses its effectiveness as a decision-making tool and guide.

Consider, for example, an agricultural equipment company; below are a set of essential variables to examine and understand the linkages between them:

- Years (Year 1, Year 2, Year 3, etc.)
- Mega Trends (world population growth, climate change, pollution, water scarcity, etc.)
- Market Segments (open land farming, Indoor farming, etc.)
- Business Drivers (productivity, sustainability, equipment reliability, etc.)
- Company Goals (revenue growth, market share, profit, etc.)
- Product Portfolio (tractors, harvesters, indoor farming equipment, etc.)
- Primary Functions (move earth, analyze soil, irrigate farm, plant seeds, harvest crop, control environment, etc.)
- Platforms (plows, mowers, sprayers, irrigation, etc.)
- Modules (blades, wheels, rakes, discs, pumps, etc.)

- Technologies (temperature and moist sensors, GPS, hydroponics, lighting, robotics, precision agriculture, etc.)
- Subsystem Supply Partners (SSP 1, SSP 2, SSP 3, etc.)
- Critical Capabilities (advanced farming, system modeling, equipment design, in situ soil analysis, precision agriculture, etc.)

The focus here is on how systems architectures can be leveraged to better capture, review and effectively use future scenarios on a roadmap.

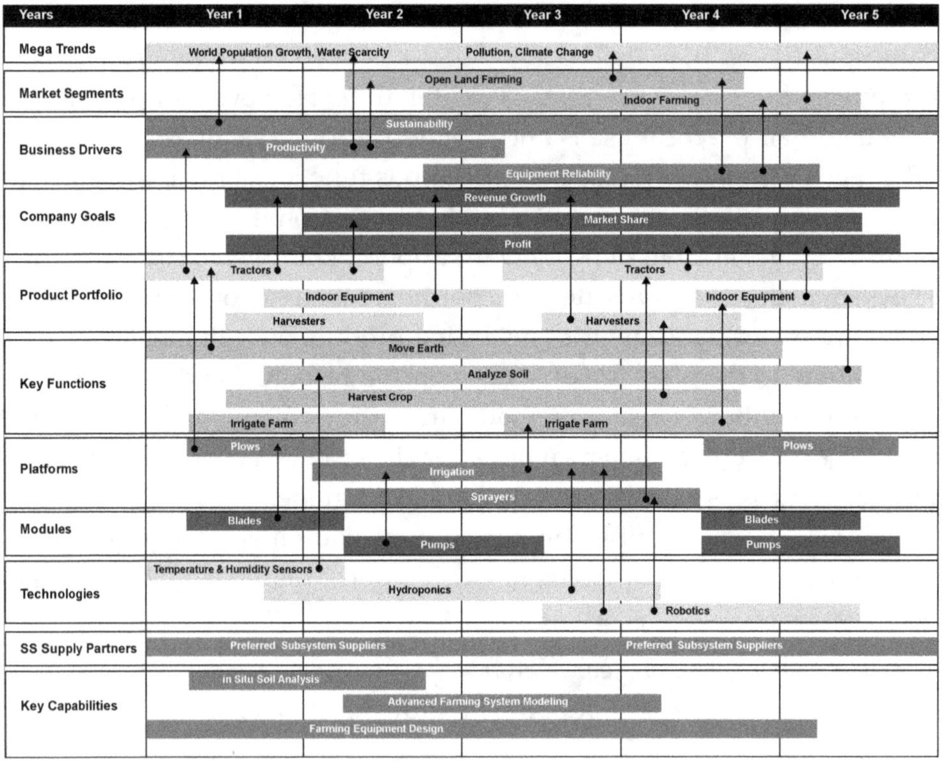

Fig. 11.3 Example of Product and Technology Roadmap

Good examples of product roadmap templates can be found in the vast literature about product development. Fig. 11.3 is an illustrative example of such a roadmap.

A few observations can be made about Fig. 11.3 and the use of a roadmap in general:

- The roadmap is a living document that should be kept up to date.

- The update of the roadmap should be triggered by events rather than by a fixed schedule. Some of the update triggers could be company goals, competitors' move that substantially impacts a market segment, new regulations, availability of new technologies, a subsystem supplier's business event, etc.
- The timelines on the roadmap should not be made rigid. Detail program plans should be kept apart from the high-level framework.
- Annual operating plans should take the slice of the year of interest rather than being completely re-created.
- The number of years covered could be less than five since the document will continue to evolve.

11.2.2 Market Requirements

All the information gained from understanding the customers and the competitive landscape leveraged to build the company's product strategy is also crucial to setting the market requirements for various products' development. Chapter 5, dedicated to systems design and mostly to requirements, has some suggestions about market requirements.

11.2.3 New Product Development

Most companies have a rigorous product development process with gate deliverables at the end of each phase or stage. The processes follow either waterfall or agile rules. But either way, they must be followed consistently to continuously improve them. As shown in Chapter 5, Section 5.3, even though companies assume that their product development process, especially for the waterfall, is linear, it is not, when considering that it starts and ends with the customers. For any product sold and any service provided, the market feedback data is precious and should be captured to improve future releases. As also stated in Chapter 5, extended programs run the risk of missing their market. The market environment is dynamic, especially for the known unmet needs, and the unknown, unmet needs comprehensible to customers. Whether the contemplated solution is incremental or radical, the needs expressed through the requirements should be validated through market feedback analysis or digital means without necessarily building hardware and testing. For longer programs, the validation should happen regularly, along with the development. Virtual laboratories, described later in Chapter 12, should be the new way for companies to fully immerse their customers into the modern world created by their products and let them digitally interact with them before making significant investment decisions. Fig. 11.4 is another reminder to move a linear product development process to one that starts and ends with the customers with the correction of the market movement.

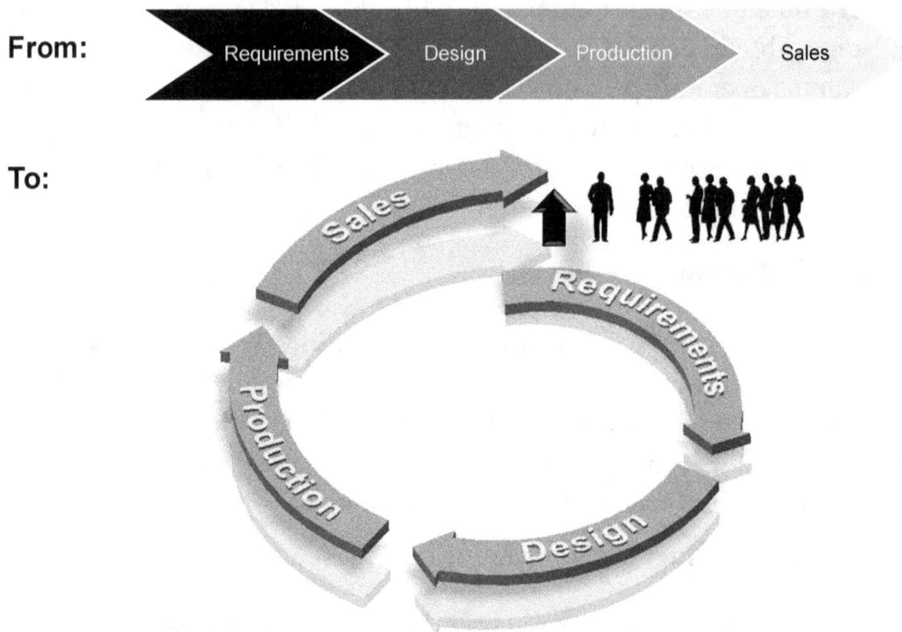

Fig. 11.4 The dynamic nature of Innovation environment

Requirements are not static; they are dynamic. After they are frozen to allow development teams to progress, regular feedback from future customers is important to understand the market movement and provide the necessary corrections. This awareness often generates tensions between the functions of product management and engineering. Product managers would like to keep tweaking product requirements even close to release to keep them as fresh as possible. Engineers would rather have them frozen early to allow them to design and test without abrupt adjustments. Such tensions are particularly evident for extended programs and less for shorter programs. Development programs of modular products offer the possibility of developing modules rather than extensive system redesigns. They are shorter by nature and are natural solutions for the tension between the functions.

Product Management is a strategic function in any company. Product managers' role in understanding the customers and the competitive landscape drives the definition of the opportunities and the right direction to develop successful products.

11.3 PRODUCT MANAGEMENT SUMMARY

The organizational function of product management guides the strategy, planning, business case, market requirements, development, release, and marketing of new products based on its knowledge of the company's customers and the competitive landscape, as expressed in the figure below.

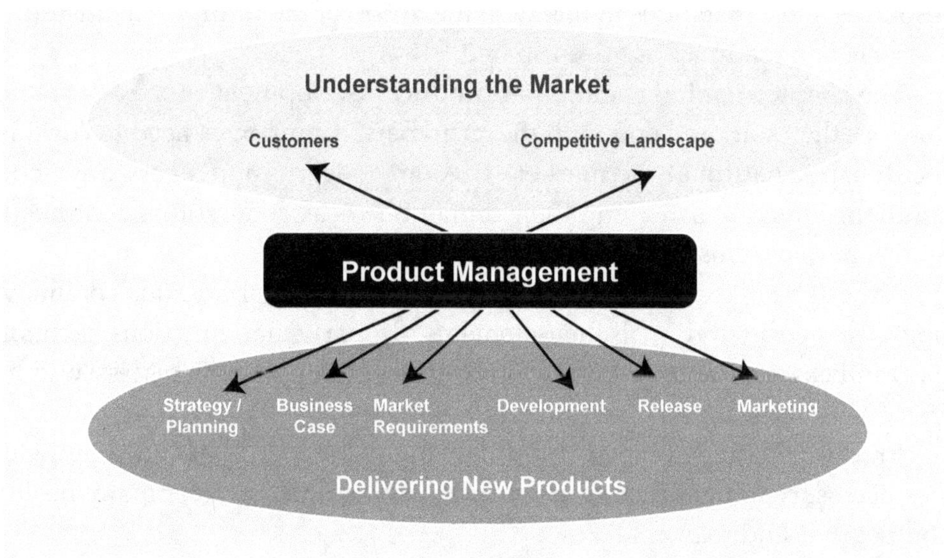

Fig. 11.1 Product Management Functions

Knowing the customers is about understanding their needs expressed by the functions performed to satisfy them. Therefore, the functional architecture is again the document of reference to be leveraged to address the unmet needs. This chapter highlights a few critical questions that the product managers, systems architects, and systems engineers should answer before starting any new product development investment.

For the competitive landscape whose understanding is crucial for positioning the company's products on the market, a system benchmarking template based on a functional analysis has been shared. The importance for companies to invest in understanding and documenting the architecture of their products is, once again, very evident.

The chapter also emphasizes some further actions to be taken to delivering products. A vital pillar to sustainable new product development is the long-term planning that links the megatrends and the business opportunity drivers to the company goals that determine the strategic choices in terms of product and technology over time. The synthesis is expressed through a product and technology roadmap that offers the framework that enables various functions or departments of a company to visualize how their different goals and, consequently, current and future activities are interdependent in support of the overall company strategic goals. They can better communicate and coordinate their actions toward a set of company performance metrics regarding product development over time. It also helps the planning of investments in resources and capabilities in line with the vision of the future. An illustrative example of a roadmap has been shared.

The chapter reminds again that product development processes are not linear as they start and end with the customers. Companies need to consider the dynamic nature of the market. Customers learn more every day about solutions aimed at satisfying their unmet needs as competitors continue to launch new systems.

Although not explicit in the chapter, the fundamental equations of innovation were considered in all their elements. In particular, functions, technologies, and execution needed for new product development are assessed in the framework facilitated by product management.

Product Management function, through *WinnoveX,*™ has the competencies necessary to be the catalyst for innovation and keep a company healthy, competitive, and relevant.

12

Systems Modeling and Simulation

I was once asked by a senior leader during a particularly important presentation about modeling and simulation a couple of questions that caught everyone by surprise. "What is modeling? And why is it different from simulation?" The two questions came out of genuine curiosity. My answers were simple:

- "Any representation of an entity, different from its natural embodiment, is a model." I answered. "We could have a human body model, the model of a plane, a hurricane, etc. Models could be physical or digital."

  Then I added:

- "A simulation is done when the model is used to show or study how the entity works; in other words, how it performs its functions. In this case, we could have a physical or digital simulation. Also, a simulation's greatest value is that it can be used to explore behaviors the system could have in certain conditions."

Simulations cover scenarios that can be difficult to execute with real systems.

When considering modeling, in general, a distinction should be made between physical models and digital models. For physical modeling, the model is physically built from various materials. A car model that is put in a wind tunnel is usually made from foam board. Medical students work on human body models before getting to learn from cadavers and then actual patients. Digital models are mathematical, logical, or process models, etc.

In the same way, we could distinguish between real and virtual simulations. In real simulations, real entities are imitating or playing actual events, such as movies, military, or firefighters exercises, etc. Virtual (digital) simulations cover either real entities interacting with virtual conditions or cases where both entities and situations are virtual. As for terminology, we will be using digital and virtual interchangeably to differentiate them from physical modeling or real simulations.

The last few decades have seen unfathomable progress in the field of computation. New capabilities have democratized the accessibility to digital modeling and simulation to design new systems. A product can now be designed and have its manufacturing, assembly, logistics, and service operations simulated before any hardware is built or purchased. A phenomenon can be simulated and reach conclusions to prevent disastrous events in the future. We live in an era where we can leverage "More Digital and Less Physical" capabilities in new systems development. Companies used to conduct several "Build and Test" cycles where a proof of concept is built after a product's conceptual design.

In some cases, several of them are built before proceeding to detail design. Then, one or several prototypes are manufactured, assembled, and tested in a loop that could take months to complete. At every stage of its development, a product could see several builds. Each comes with serious investments of precious resources. And sometimes, the cost to build and test a system several times before its release can be simply too high. The last decades of the 20th century and now the 21st, have allowed companies to cut development cost and time by using more digital and less physical approaches to develop new systems.

Any modeling, but mostly digital modeling, starts from understanding the system of interest. The subject to model could be a physical entity, a phenomenon, or even a process. The first goal is to represent it as truthful as possible. Failing to do so would only give results that would lead to false conclusions. Therefore, every effort must be made toward the fidelity of models, which need to be continuously updated as more details or data become available. They also need to be validated before their use to reach actionable conclusions.

The purpose of this chapter is to explore some of the opportunities offered by digital modeling and simulation in support of WinnoveX™ activities. So, our emphasis will be on the applications in a modular architecture environment. Such an environment enhances the capabilities already massive of new computation techniques.

12.1 PHYSICAL TESTING VS. VIRTUAL LABS

12.1.1 Physical Testing

A product that is well designed with a clear architecture made of subsystems, modules, components, and parts, could execute physical testing at each level. System-level testing is mostly around performance, reliability, integration, regulatory, and a few other tests to prove that it is safe for general use. Inside a company, testing would validate that the product meets market requirements (customer needs) and is ready for sales. But depending on the industry and the criticality of the applications, testing as one of the requirements verification methods, can go deeper into the system's elements. The International Council on Systems Engineering (INCOSE) describes other ways of performing requirements verification.1 The Council highlighted verification through inspection, analysis (modeling & simulation, analogy or similarity), demonstration, test, and sampling.

Subsystems are often tested when they are safety-related, and their failure could be catastrophic to the whole system. The same can be said about modules, components, and even parts testing. As shown in Fig. 12.1, the overall testing cost can increase rather rapidly when testing is extended beyond the system to its lower level constituents from A to E. But extending testing to subsystems, modules, components, or parts has several advantages:

- It allows to verify their compliance with the requirements allocated to them,
- It gives a more reliable reading of the level of safety and normal behavior of the lower-level elements,
- It enables the verification of some redundant safety measures at the right level,
- It limits the system level testing to its general behavior, mostly driven by the interaction of its modules highlighting any emergent undesired behaviors,

1 INCOSE, *Systems Engineering Handbook*, Fourth Edition, Wiley, Section 4.9.2.2 p.86

- Defects are caught and fixed at the proper level.

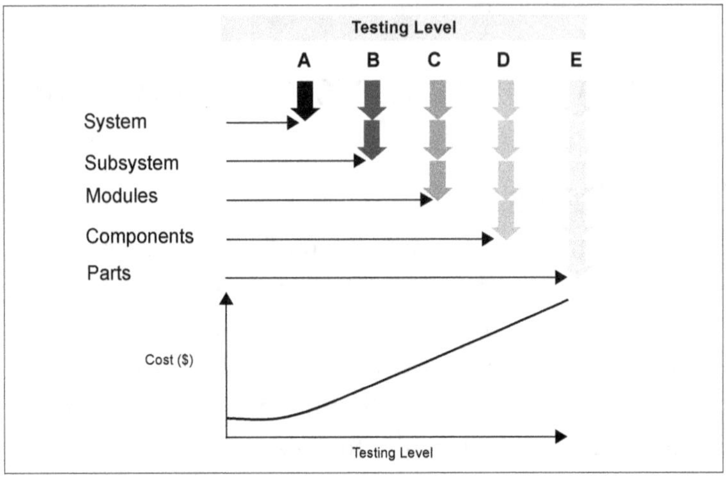

Fig. 12.1 System-level vs. Component or Module-level testing cost

Even though testing a system's elements bottom-up may be costly, it can be even more expensive to find and fix a defect at the system level rather than at the component or module level. The cost of fixing defects is inversely proportional to the testing extension to various levels. It will cost more to find and fix defects at system level A than those found at part level E.

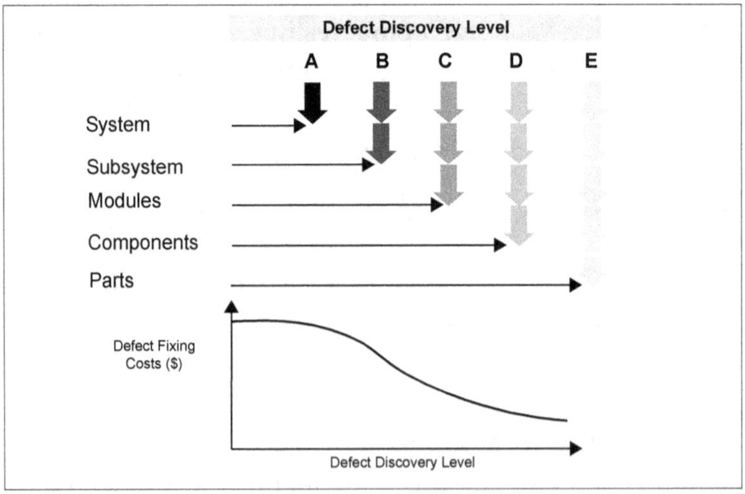

Fig. 12.2 System-level vs. Component or Module-level defect finding and fixing cost

In addition to the financial cost, the time cost is similarly inverse. This observation is expressed in Fig. 12.2, which shows how defects finding and fixing cost decreases as fewer elements are integrated down to the part level.

Nobody can refute the value of testing. It is so important that some manufacturers perform it themselves in house (in their own labs) or engage specialized laboratories and government agencies to carry it out for them. Some countries or group of countries like the United States, the European Union, and others, have created special certifications, marked by labels, to prove that the released products are suitable for their citizens.

The graphs in Figs. 12.1 and 12.2 highlight two main aspects of physical testing relative to cost. Testing every module and subsystem in addition to the system can be costly. But not testing them right can be more expensive if defects are found and fixed at the system level. Testing, however, does not need to be all physical. The progress made in computation and the digital environment, in general, has opened new opportunities.

12.1.2 Virtual Labs

One of the most significant modeling and simulation applications in the industry, especially for new product development, is to reduce physical testing. Let's consider testing in the digital or virtual world. We will call digital testing any simulation made from a digital model where the system or its subsystems, modules, or even components are digitally put in their working conditions to study their behaviors in compliance with their requirements. In the process, during their operational scenarios' review, the team can make new assumptions and analyze their response, as potential failure modes or application opportunities emerge. Digital testing presents the opportunity to reproduce scenarios that would be too expensive or too difficult or unsafe to obtain through physical means.

The future, which is already here, is seeing digital testing applications through modeling and simulation in setups that are practically full-fledged virtual labs. Like any lab that tests physical (hardware-based) samples, digital environments are growing more sophisticated to represent practical labs. In this case, all the tested models are digital. Even though they do not and will not eliminate physical testing due to some verification and validation tests on actual units, they can dramatically cut down its cost and time. Certification bodies worldwide still require physical testing that they can witness, even though virtual labs are starting to make their way toward that microcosm. Digital testing in virtual labs offers quite a few additional advantages.

We can observe:

- The possibility to digitally test (model and simulate) each module, including its components and parts, and even test some components and parts separately from their assemblies.
- Any combination of system elements that communicate with one another can further be tested to understand their interactions.
- The signature of each module being identified to monitor its health over time. For example, a module with rotating components and parts can have sound pressure (noise) levels at specific frequencies when they are healthy. If the noise level is different at such frequencies, the system sets to activate an alert mode, sends a warning. If the parameters get worse, it generates an alarm mode, and asks for an intervention. New conditions are simulated in "What If's" scenarios. Such an achievement allows exploring the system's capabilities to validate claims and better position the product in the marketplace.
- That newly designed modules can be simulated as many times as possible in a loop that does not carry safety risks and at a much less cost before any unit is built for final validation testing.

Virtual labs offer the opportunity to test any system constituent at any level and reduce the risk of finding defects late in developing a product, which comes with high cost to fix. The question is how such an approach is made practical without becoming a costly undertaking.

12.2 VIRTUAL LABS POWERED BY "MODULAR SYSTEM ARCHITECTURE"

The observations of Fig. 12.1 have highlighted a few opportunities that enhance the advantages of "expanded level" testing, when done digitally, without carrying the cost that a physical lab would require. Virtual testing allows a significant number of scenarios, including those with less probability to happen but would have disastrous consequences if the system fails. With the adoption of modular product architecture, each module or subsystem can be simulated. If a component is critical for the system's safety and performance, its modeling and simulation can be extended to include tests that demonstrate its safety and functionality.

There are a few steps to fully implement a system and its critical elements' modeling and simulation. These steps are reviewed below as they apply to the bicycle portfolio example used in Chapter 3, represented again for convenience.

A. Functional Architecture

- Define the functional architecture of the system or portfolio,

B. Requirements

- Define the market requirements,
- Translate the market requirements into system requirements with an emphasis on the acceptance metrics to meet,
- Allocate the system requirements to the subsystems,
- Allocate the subsystem requirements to the modules,

C. Verification and Validation

- Create the Verification and Validation Plan and execute it at all the appropriate levels of the system.

A brief description of each of the steps follows and helps grasp them better; their details, however, can be found in Chapter 5 and Appendix C about systems design and particularly about requirements.

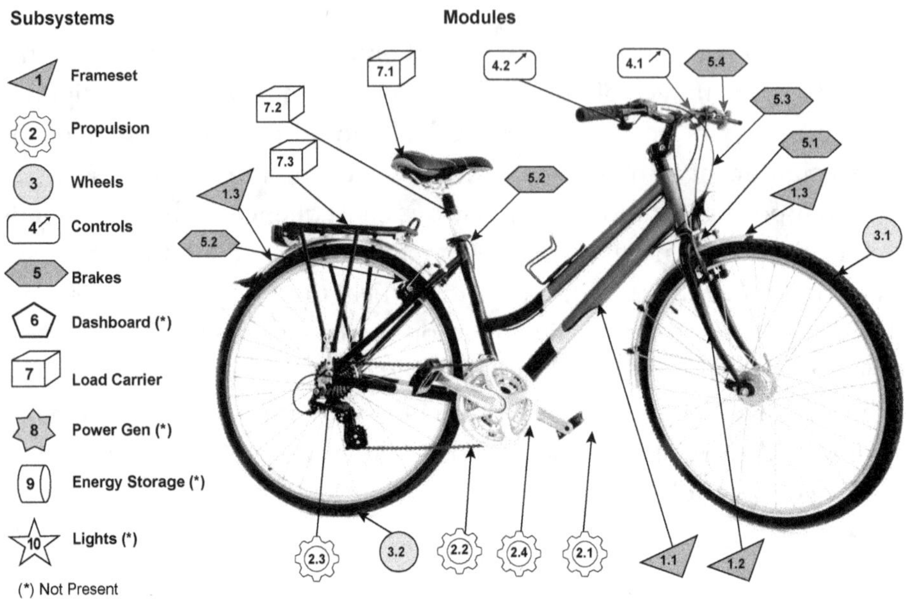

Fig. 3.8 City bike subsystems and modules

A. Functional Architecture

The system or portfolio of systems' functional architecture is, once again, the starting point of the modeling and simulation journey.

The bicycles' portfolio's functional architecture, including the one represented above and other design and configurations (mountain bikes, race bikes, motorized and electrical bikes), is shown in Fig. 3.12, also brought here for convenience.

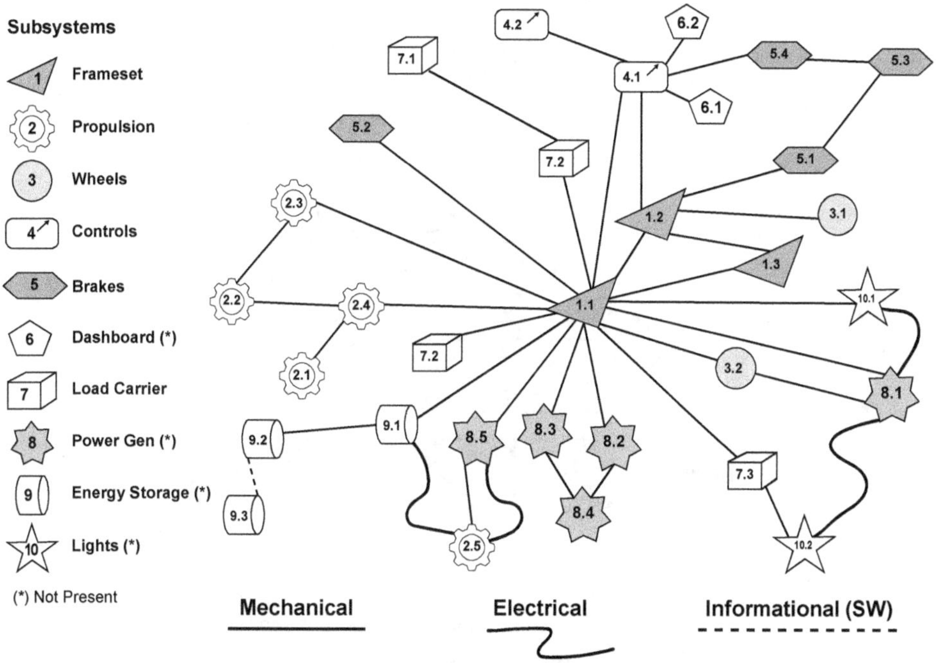

Fig. 3.12 Functional architecture of configurable bikes

B. Requirements

As stated in various sections of the book, the functional architecture investment should be made outside of any specific product update or new development. Such a choice allows the company to consider all its offerings and identify the functions they perform to set the architecture that encompasses the products or systems that perform the same functions. As explained in Chapter 5 and Appendix C, market requirements should be defined under product management's leadership, for any specific development. The product development team's technical members should translate the requirements written in a marketing language into more technical terms before allocating them to the subsystems and their modules and any platform that may have been defined. Any critical component should also have its requirements unambiguously defined. Fig. 12.3 below shows the modules of the bike example, grouped into their subsystems.

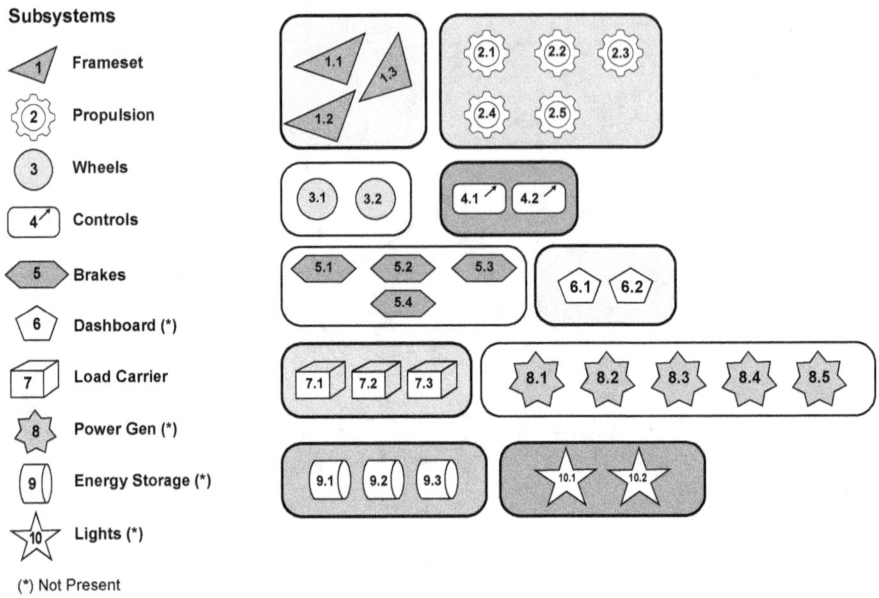

Fig. 12.3 Bike portfolio subsystems and modules

The requirements are then allocated to the subsystems and the modules, as shown in Fig. 12.4.

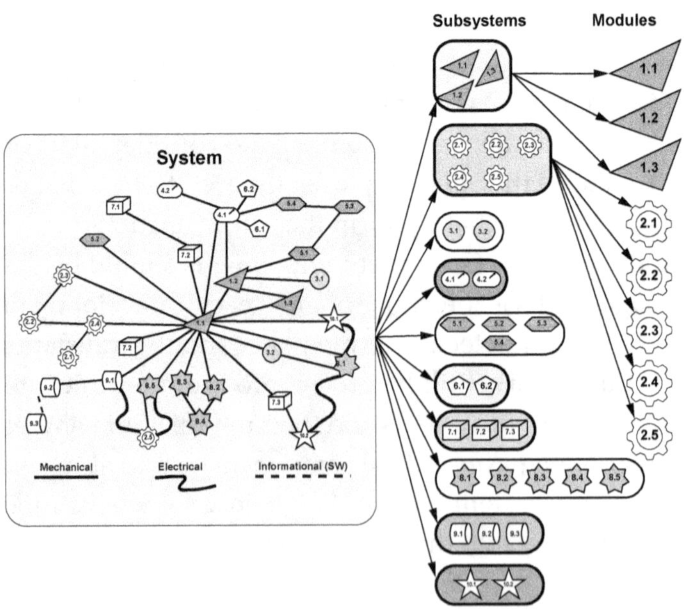

Fig. 12.4 Requirements allocated to subsystems and modules

C. Verification and Validation

As described in section 12.1, testing is necessary to confirm that the system meets its requirements and, more importantly, the customers' needs. It is also essential to avoid field failures that could come with a high warranty cost, product recall, and company brand damage.

At the system level, both verification and validation are shown in Fig. 12.5, where customers' identified needs are translated into system requirements with some of its most important sections. Each of the requirements' sections' metrics needs to be verified through at least one of the methods mentioned in section 12.1 to make the system ready for its validation process. In support of the final system-level phase, the verification process should have been carried out at the subsystem, module, component, and even at the part level, as shown in Fig. 12.6.

The steps are less expensive when leveraging digital testing compared to physical tests. The investments to build good modeling and simulation capabilities should be made with senior leadership support as part of the company's strategic choices.

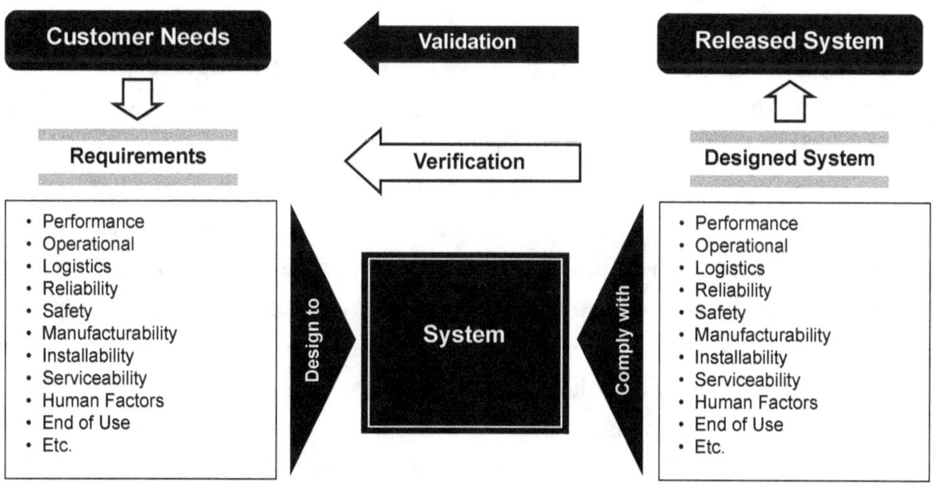

Fig. 12.5 System-level verification and validation

While the System requirements are allocated top-down to the subsystems, modules, and components, the digital verification happens bottom up back to the system, as shown in Fig. 12.6. The system-level process was represented in Fig. 12.5.

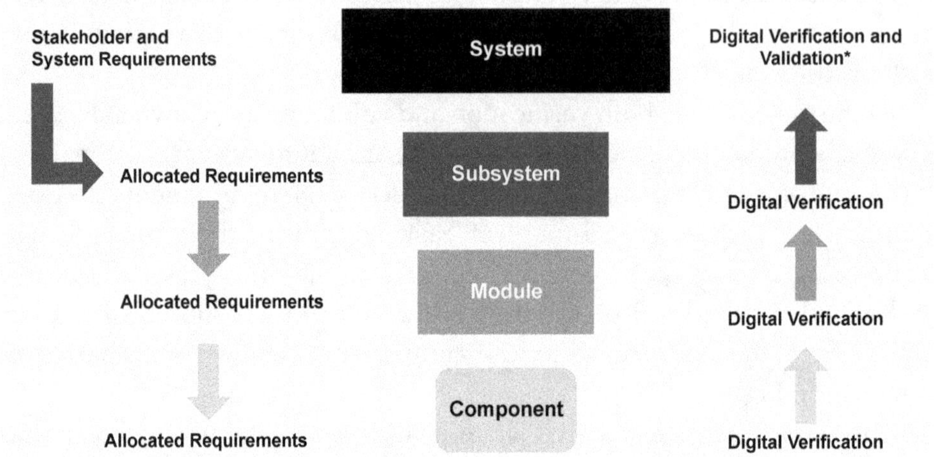

Fig. 12.6 Digital verification of allocated requirements

(*) Also, limited physical validation at the system or product level to demonstrate compliance to critical safety, regulatory or performance requirements.

Companies should make modules' verification, through digital testing, part of their design process. There are good software packages available on the market that enable the modeling and simulation of system modules with their components. Physics-based modeling has made remarkable progress in the last three decades. With one output becoming the next assessment's input and sometimes requiring a new model of its own, series of discrete analyses are now seeing multi-physics applications working off the same model and often using the same computer-aided design (CAD) model. These new possibilities enable the designers to verify that a module meets its requirements. They can also simulate its behavior within the system through its input from a set of modules and its output toward other modules. Now, the possibility to leverage the abundance of data collected from an installed base to understand the behavior of a module, a subsystem, or a system being designed, is fast-growing and opens new doors for the future.

With digital verification coupled with system design, modules with their components and parts become strategic assemblies leveraged to build platforms, subsystems, and ultimately systems. While they are being designed, the modules can be verified in a productive loop that would, otherwise, be expensive and time-consuming in the physical world. Fig. 12.7 shows the

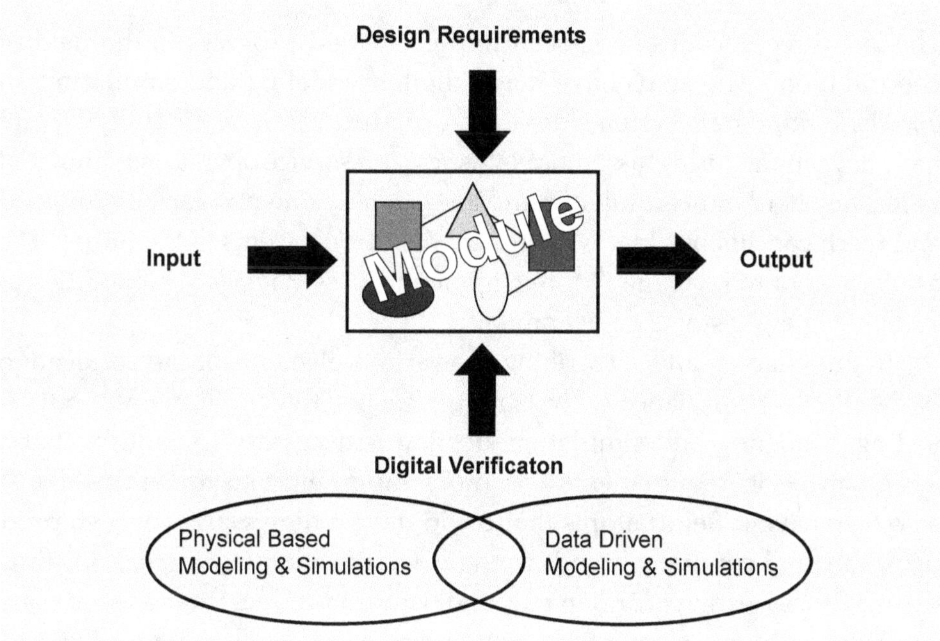

Fig. 12.7 Digital verification of a module with its components

context of module design and verification.

Once a newly designed module is digitally verified, it can be plugged into the system model to understand its impact on the unit. Module models can be selected from a library and be plugged into a digital system in a virtual lab with possibilities that will only continue to grow. Coupled with technologies such as virtual or augmented reality, customers can now experience their future product even before making any part. They can provide instant feedback making it more impactful for the designers. Such an approach is especially crucial for radical solutions for unknown, unmet needs incomprehensible to the customers when expressed, as presented in Chapter 6.

12.3 SYSTEMS MODELING AND SIMULATION SUMMARY

The last three decades have seen unprecedented progress in the field of computation. The steps have made digital modeling and simulation an excellent aid to new systems design. A product can now be designed and have its manufacturing, assembly, logistics, and service operations simulated before any hardware is built or purchased. A phenomenon can be simulated and reach conclusions that will prevent disastrous events in the future. We are in the era to leverage "More Digital and Less Physical" capabilities to accelerate new systems development.

In this chapter, after describing the various elements of the verification of a system's compliance with its requirements, the emphasis was put on testing, modeling and simulation. Testing is necessary to confirm that a system meets its requirements and, more importantly, customers' needs. It also helps avoid field failures that could have a high warranty cost, product recall, and company brand damage. Certification testing and, in some cases, witness testing are often required before a product is released for sale. However, not every element of a system can be tested. As shown in Fig. 12.1, development cost can quickly rise as more system constituents are tested at various levels.

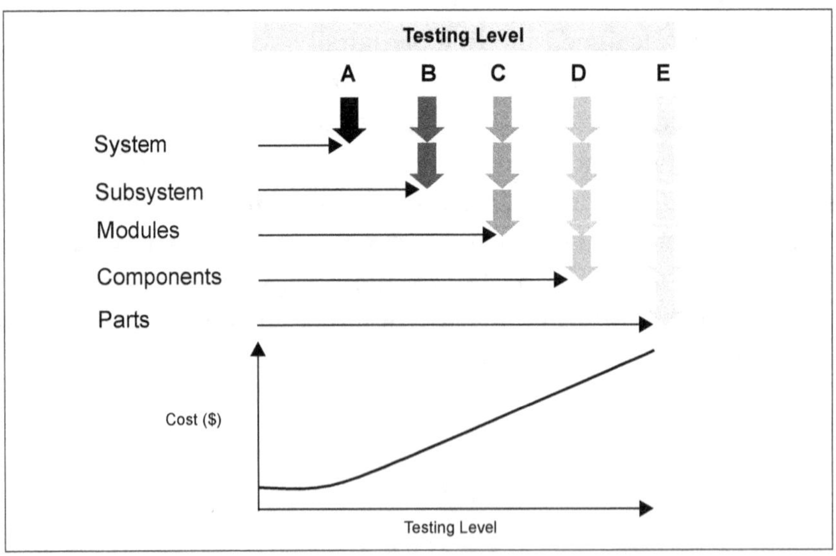

Fig. 12.1 System-level vs. Component or Module-level testing cost

It was also shown in Fig. 12.2 how, by not testing at the proper levels, the cost of finding and fixing defects at the system level could be significantly higher than acting at components or modules levels.

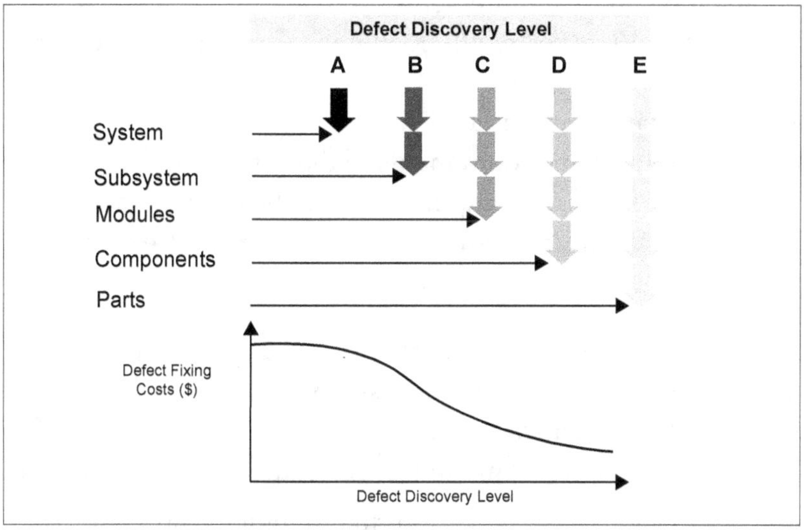

Fig. 12.2 System-level vs. Component or Module-level defect finding and fixing cost

The graphs highlight two main aspects of physical testing relative to cost. Testing every module, subsystem, and system can be costly, but not testing at various levels can be more expensive if defects were found and fixed at the system level. With computation progress, by leveraging digital testing, a system's constituents can be designed and tested in full-fledged virtual labs. Every module and subsystem can be digitally tested individually, and most defects can be fixed before making any part. The convergence of physics-based modeling and data-driven modeling is making it possible to model and simulate whole systems. Once the system's digital model is created, it can continuously be improved based on the installed base's data. Then, any module being designed can be digitally replaced in the system model to assess its impact on the whole. Physical testing would mostly be limited to specific investigations or at the system level for overall performance validation, customer acceptance testing or regulatory compliance testing, and any particular case with the new capabilities.

A few steps needed to fully leverage a virtual lab have been reviewed. They include,

- Defining the functional architecture of the system or the whole portfolio,
- Defining the market requirements,
- Translating the market requirements into system requirements with an emphasis
- On the acceptance metrics to be met,
- Allocating the system requirements to the subsystems,
- Allocating the subsystem requirements to the modules and,
- Creating the Verification and Validation Plan and execute it at all the appropriate levels

These are the key steps that enable the creation of a product portfolio's digital library with the systems, subsystems, modules, and components and a virtual laboratory in which products can be digitally tested. Such an investment would open, for any company, a few leading-edge opportunities:

- Product time to market reduction due to digital testing of system constituents,
- Product design and development project cost reduction (reduction of the number of physical builds) and,
- Technical risk reduction due to both module and system simulations during design.

Modeling and simulation capabilities offer excellent tools to companies on their way to winning innovation excellence or *WinnoveX.*™

13

Organization Design and Human Dynamics

Aristotle, the Greek philosopher, recognized as the founder of political science, in his famous expression "Zoon Politikon", characterized humans as social animals due to our need to gather, communicate, and organize our lives as groups. Humans live in groups, from a family as the nucleus of any society to large populations in cities and countries of our current modern era. Throughout our history, human scientists have dedicated their careers to studying human beings. Anthropologists, sociologists, psychologists, and others have tirelessly explored multiple aspects of our lives, as individuals and groups, and described their complexity and their richness as a system called society. The consideration made by Emile Durkheim, a French sociologist,[1] to view society as a whole entity more than the sum of its parts, is remarkably interesting as it directly compares society to a system. Also, such a comparison makes sense from several viewpoints:

- Like a system, society is made of elements that, as a whole, delivers values that no single constituent would achieve alone.
- A society performs high-level functions that lead to "Improve Lives." The need to stay together is justified by the security and the ability to achieve more as a group than as individuals.
- Whether it is a subgroup or an individual, each element performs a function or a set of functions that contribute to the whole.
- The exercise of such functions produces the goods and services human society leverages to meet its own needs.

Beyond a family where the members' selection happens following rules governed by the complexities of "Love", which science is not well equipped

1 *Internet Encyclopedia of Philosophy*, www.iep.utm.edu

to fully understand and replicate, most other social gatherings are intentional. The assumption is that, beyond certain natural groups, most of the others are formed for a purpose. Assuming that such human-made groups function as systems, it makes sense to think about their optimization, from their design to their execution. This chapter's main objective is dedicated to what could be characterized as optimization of "human systems" design and execution. We will review how *WinnoveX*™ applies to human systems.

13.1 ORGANIZATION DESIGN

Everything outside of a natural phenomenon and environment is created by design by human systems; this means that there was a scope and a plan to its achievement through means that were thought ahead of the execution. A few examples to clarify the statements made above would include:

- **An army:** armies are formed to defend a community's interests and possessions against another society intended to usurp them. From the dawn of civilization, armies as human systems are made of people with certain physical conditions with the means and strategy to keep their enemies away. They are structured in subgroups, in a hierarchy, where each team has precise functions to perform. The purpose function would be to "Protect the Community." Such a high-level function can be decomposed into, fight enemy, hold ground, defeat enemy, minimize casualties, etc. Even the strategy is defined with functions to perform. Each of the functions is further decomposed and assigned to modules of the whole system to fulfill. The means used are all concerning their functions. The system executes its mission every time the triggers of its exercise are activated.
- **A company**: companies generally start from founders who identify an opportunity through solving a problem by providing a solution to an unmet need. The founder can express such a solution that could be a product, a service, a process, or an organization by performing the main purpose function. In 1853, Elisha Graves Otis showed the world he could safely stop a hoisting platform if its ropes were severed. He invented a safe way to prevent it from falling. Solving that unmet need

gave rise to modern cities by enabling safe elevators. Fred McKinley Jones solved the unmet need to transport perishable goods by inventing a portable air-cooling unit installed on vehicles. Like Otis and Jones, company founders find ways to meet unmet needs expressed through functions to be performed. The purpose functions can be decomposed top-down into many other functions that enable the mission to be realized.

13.1.1 Functional Organization Design

Like a system, an organization needs to be designed based on its purpose function that is then decomposed into many other functions that make the organization successful. Also, like any system, the functional requirements of an organization need to be carefully defined and used to design its structure with the roles and responsibilities. Interestingly, the so-called roles are the functions to be performed by those responsible and in the relative positions. Before getting practical with some organization design, it is worth making a few observations:

1. A company that invests in the overall functional requirements of its organization can create an architecture as an organization structure with the right positions and the minimum requirements for the profiles of people needed, not only to fill a specific role but to fit with other resources for the best performance of the whole.
2. Any organization as a human system can be reviewed and reshaped as the purpose function or functions change.
3. Unlike other systems, human systems are not deterministic. We cannot push a given button and expect to get the same exact "response" every single time from a social system, but that is what makes an organization a far more important investment than any other one in a company. This adaptability and capability of tailored responses place people at the core of every organization and its resources. Human systems are adaptive to new realities when clear strategic directions are provided. So, the time, energy, and resources invested in people to train and grow them have an invaluable return that can propel a company, making it competitive and innovative.
4. Once the roles are defined, and the organization architecture is set, it

should not be static but reviewed regularly and kept up to date in line with the company's mission to serve its customers. If the mission or missions change, the functional requirements should be reviewed and updated. Such a move may, in turn, require an adjustment to the roles and responsibilities of the various positions. Some people may need to be re-trained to accomplish the new mission and the new or updated set of functions they need to perform.

5. When a resource leaves a company, it gives the opportunity to look for talents that would better perform the existing functions. It may also be the time to review and update the functions and change the profile to match them. Unfortunately, the leaders look to fill a position with the same profile as the previous person that occupied it without reviewing the requirements for a better performance of the existing functions or review the functions to update the profile. Such a practice should be discouraged, and efforts should be made to push the organization to review its functions and decide if any change is necessary. By doing so, the job description would define the requirements for the new hiring.

After these few observations, we can explore how an organization should practically be designed based on its defined functions (roles).

Chapter 2 described how system analysis should be performed. Fig. 2.2 recalled below, showed how the main purpose function could be decomposed into lower-level functions. The figure was a generic functional decomposition. It would be better understood relative to human systems with an example.

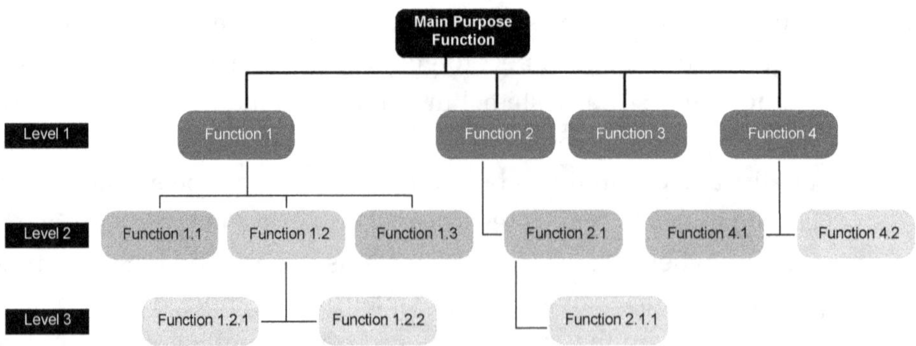

Fig. 2.2 Functional Decomposition

A division of an established fortune 250 company decided to reorganize its engineering to better serve its customers with new technologies, new products, and the continuous upgrade of existing products. So, its senior leaders tasked the engineering leader to reorganize his team involved in supporting several product lines which covered other locations worldwide. In this case, the main purpose function for the team was to "Drive Product Leadership." The new team's mission was to deliver products perceived by the customers from the markets the company served as superior to any other offering. The engineering leader knew that he could not operate in a vacuum. Therefore, he and his colleagues from product management, human resources, operations, and finance formed a team. Spending a couple of days offsite, the team reviewed the main purpose function and decomposed it into the functions below:

- Understand strategy
- Execute strategy
- Develop new technologies
    - Review strategy
    - Select technologies
    - Engage technology partners
    - Grow technologies
- Develop new products
    - Manage requirements
    - Analyze systems
    - Update architecture
    - Simulate systems
    - Manage subsystems
    - Design core modules
    - Design non-core modules
    - Qualify modules
    - Integrate systems
    - Test systems
- Sustain existing products
    - Review field reports

    - Resolve field issues
    - Review cost
    - Reduce cost

- Support sales
    - Review specs
    - Provide deviations
    - Resolve issues

- Support supply chain (including manufacturing)
    - Support Bill of Materials
    - Review deviations
    - Review processes
    - Review design
    - Align design

- Support Field
    - Review processes
    - Review design
    - Align design

- Manage projects
    - Plan projects
    - Track progress
        - Track completion of work packages
        - Cost
        - Track schedule

- Provide status reports
- Resolve issues

The decomposed functions were then summarized in the functional structure in Fig. 13.1

The engineering leader and his team leveraged their internal customers (product management, supply chain, manufacturing, field operation, and service operations) and support organizations (human resources and finance) to better understand the company's strategy to execute. Then, all the engineering functions were captured. They further decomposed them to get the roles' details closer to the people and positions that will execute them. Several discussions took place around the functional organization, and everyone was aligned. Only then, the physical organization design started.

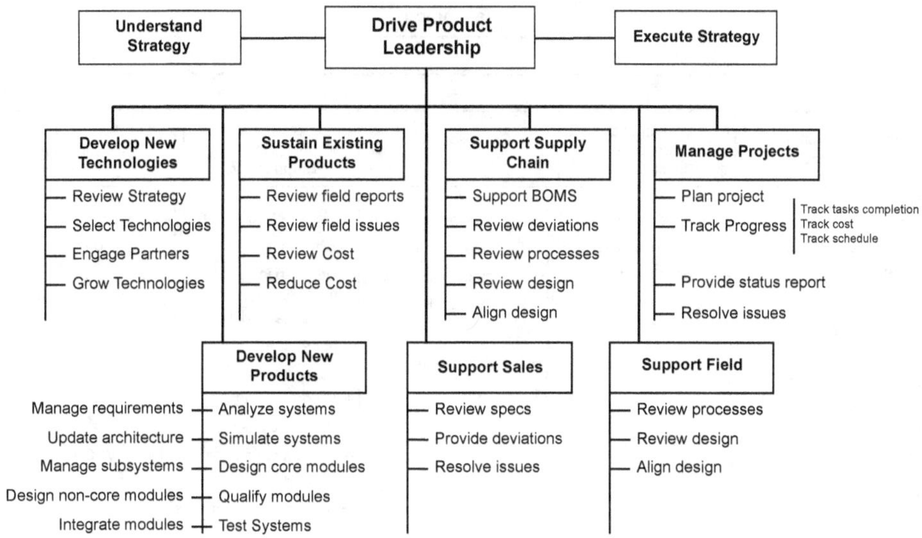

Fig. 13.1 Example of an engineering (organization) functions structure

13.1.2 Physical Organization Design

Once the functional structure was defined, the functions were grouped according to a set of criteria that would maximize performance to drive product leadership as expressed and expected by the company's leaders. The main criteria were:

1. Minimize footprint
2. Maximize flexibility to meet customer demand
3. Accelerate innovations
4. Minimize time to market (reduce product development cycle-time)
5. Provide global coverage
6. Maximize competencies
7. Ease interactions with internal and external stakeholders
8. Maximize talents attraction
9. Minimize overlapping of roles and responsibilities

Following the criteria, the functions were reviewed, and the functional organization was created, as shown in Fig. 13.2.

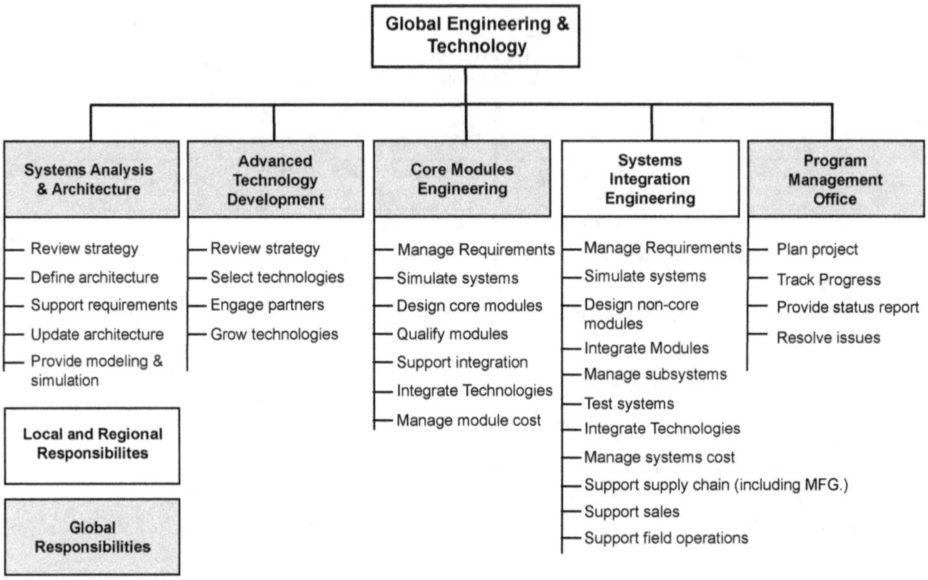

Fig. 13.2 Example of Engineering Functional Organization

- A system-level group was created to oversee all the products' architecture and provide an analysis of all systems in support of the design teams. This group defines all the digital models used to virtually simulate the systems functionally with their manufacturing and assembly processes. The team has global responsibilities and defines any platform needed.
- An advanced technology group was created to leverage the technology sources to keep the company on the leading edge. This group has global responsibilities and works with the design teams to anticipate their technology needs and facilitate technology integration. They oversee the definition and maintenance of the technology roadmaps.
- The core modules have an organization that designs modules covering global needs in anticipation of systems integration. They work closely with systems and technology teams as well as providing support to their modules' integration.

- Every region has a systems integration engineering group responsible for designing products for their market by developing the non-core modules locally for their regions and leveraging the global core modules and platforms. They would work closely with their sales, manufacturing, and field operations teams to meet their regional customers' needs.
- The program management office's responsibility is to make sure projects meet the optimum balance of requirements (quality), budget, and schedule.

The Systems Integration Engineering (SIE) teams are the largest due to their local presence in multiple places.

The support teams are from human resources management and finance. They are matrixed to support engineering.

Once the functional organization is clearly defined with the internal stakeholders' support, the physical organization that will embody the functions is decided. Fig. 13.3 is the physical organization of the functional defined by the previous figure, Fig. 13.2.

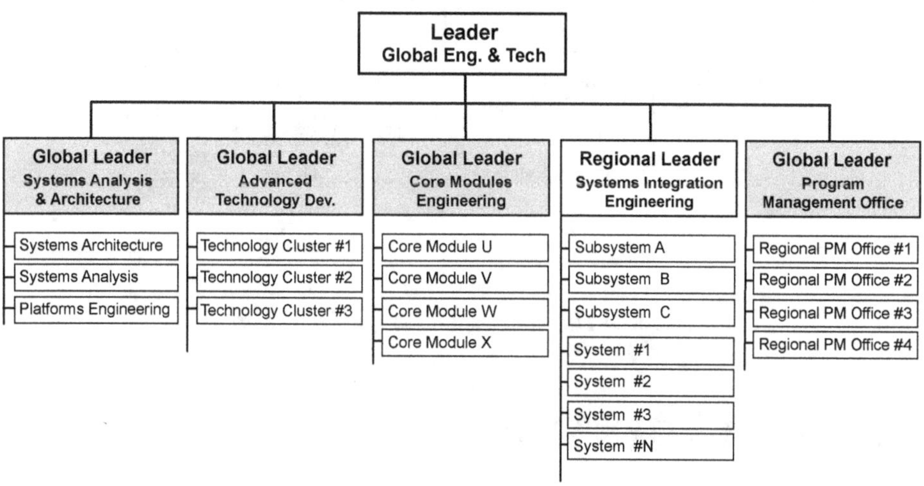

Fig. 13.3 Example of Engineering Physical Organization

Every position of the physical organization requires specific competencies and experiences. Therefore, the candidates' profiles for any of the positions and teams are defined based on the functions and skills needed to fulfill them.

The example used to illustrate the steps needed to design an organization as a human system is the same for any company. First, the functions must be defined for the team's main purpose. Then the functional organization is set. In the end, the physical organization is decided before the skills and experiences required for each position are defined. Fig. 13.4 shows the framework of the steps to human-system design.

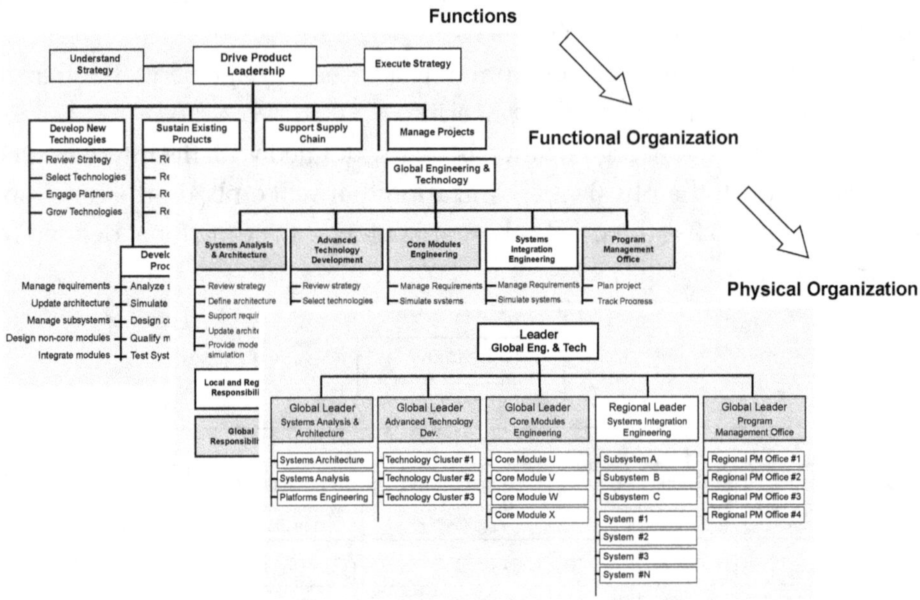

Fig. 13.4 Human System Design Framework

13.2 HUMAN DYNAMICS

Any human-made system can be complex when it must perform a large number of functions or sets of functions. Human systems become more complex due to the added dimensions of culture and individual sensitivities. Machines do not have motivations, but people do. Individuals and teams must be motivated to be engaged, and engagement is the key to success. Since human systems are the ones that make any other system work, their health determines the health of all the systems a company creates or manages. If one team is motivated and the others are not, the overall organization will not be healthy, and results will show it. When all the organization teams are engaged, their interactions are real energy multipliers, but the opposite is also true.

Attention to human dynamics to keep employees engaged is the most critical investment of any company's leadership team. Failure to realize it is worse than ignoring fractures in a dynamic load-bearing structure. It will not be long before it collapses. Leaders know it. They invest their time and energy in ensuring the primary purpose function, or functions of an organization are clear to all the people managers first and ultimately to the employees across all the company's organizations.

Figs 13.5a and 13.5b show how everybody should understand and work for the customers. In the company, starting from the external customer-facing employees who should be the most important internal customers, everyone in the hierarchy is the customer of the next layer up to the CEO as the supplier of all. In addition, Fig. 13.5b shows that not only everyone has internal customers, but they also need to have their eyes on the external customers no matter their internal position.

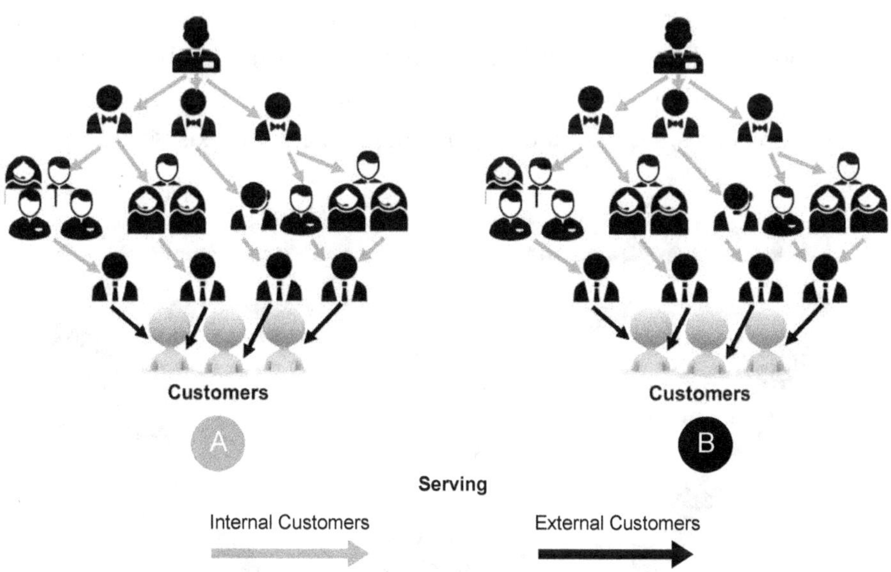

Figs. 13.5 a and b Customers Hierarchy

Such a view creates a whole set of dynamics that leadership teams need to know.

- Employees must be empowered to make decisions and prioritize the external customers' interests over any internal customer.
- Leadership needs to create a culture where employees can prioritize the appropriate customer (internal or external) in the proper time/condition.
- Leadership needs to create an atmosphere where an employee that is interacting with an external customer has the support of the whole system to represent the company and deliver results on its behalf no matter his or her role.

13.3 HUMAN SYSTEM GOALS DEPLOYMENT

As a human system, an organization can also greatly benefit from the systems engineering "V" principle to set and cascade goals. Fig. 13.6 shows how this could be applied. The left-hand side shows company hierarchy. The goals cascading mechanism should be the same as a functional decomposition (Chapter 2) and allocation (Appendix C) to a system's constituents. The right side mapped the corresponding goals from individual employees to the company strategic level. The goals should be verified to ensure they build up to fulfill the company's strategy and satisfy the stakeholders' needs (competitive quality products, processes, services and organization, financial needs, social needs, etc.).

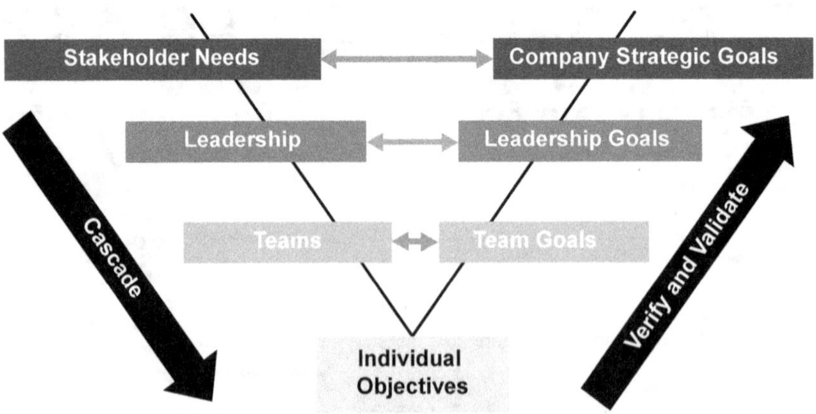

Fig. 13.6 Human system goals deployment

13.4 ORGANIZATION DESIGN AND HUMAN DYNAMICS SUMMARY

Every organization is a human system and can be governed by the same principles as any system from its design to its execution and deployment. The steps are practically the same.

- Leadership must clearly define the main purpose function or set of functions of the organization being setup.
- Then, it is decomposed into all the functions that need to be performed to make it happen.
- The functions are grouped according to a set of criteria defined by the company's internal stakeholders, and the functional structure is created.
- The next step is to create a functional organization that should be reviewed, amended, and approved by the internal stakeholders.
- The functional organization drives the physical organization, followed by the definition of roles and responsibilities, and the skills and experience required to fulfill them.
- Only then can candidates be mapped to the positions.

Fig. 13.4 summarizes the steps needed to design an efficient human

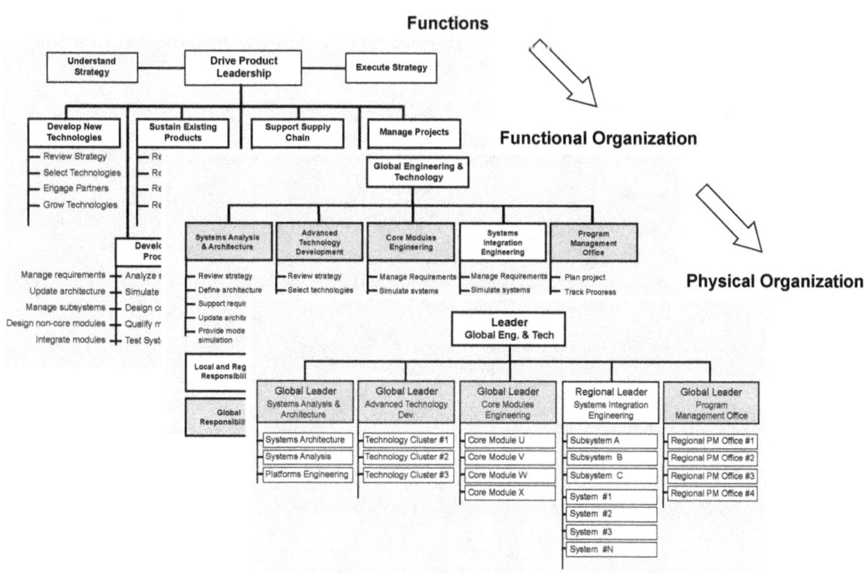

Fig. 13.4 Human System Design Framework

system.

Following the organization's design is understanding and managing its dynamics which can add more complexity to human systems than artificial systems. Overall, an organization design requires the same rigor exercised in the design of products.

Human systems are more complex and critical than machines due to their non-deterministic nature, but fortunately they are more adaptive than machines. When they are well motivated, they can quickly assess a change and provide the best response. Such a capability makes an employee organization the most important assets of any company. No organization design should be complete without the careful assessment of the team's mission and all the functions it needs to perform, followed by each individual's functions in terms of roles and responsibilities. It is an investment with a return that can propel any company to a leading position in its field and its market. The results would be seen through:

- Superior customer service,
- Product quality,
- Reduction in product cost
- Reduction in Selling, General and Administrative (SG&A) expenses,
- On-time delivery,
- Increase in the top and bottom lines of the business
- Additional generation of ideas to improve the business since team members would feel empowered to manage the company like their own.

14

A Few 21st Century Technologies and Innovations

In Chapter 6, the fundamental equations of innovation or *WinnoveX*™ equations have three key elements:

- The functions of the system,
- The technologies selected to perform them and,
- The execution of the applications

In the previous chapters of the book, we explored the identification process of an entity's functions from different methods, and their combination with technologies to build innovative systems. Execution comes from the rigor by which systems development best practices are implemented. This chapter is dedicated to a few technologies and innovations of the 21st century that can enhance the ability to innovate and offer mass customized solutions. The motivation is to shed light on how the technology element of *WinnoveX*™ gives "life" to functions in modules that satisfy conscious and unconscious customer unmet needs. The list could be much longer and vary depending on the assessor. This handful of innovations and technologies are picked because of the significant impact they are having and will continue to have on changing how systems are designed, manufactured, delivered, and serviced. We shall see how *WinnoveX*™ equations apply to them:

(1) Functions + Technology = Innovation

(2) (Functions + Technology) x Execution = Winning Innovation

14.1 TECHNOLOGIES

History has taught us that technologies played the key part in various industrial revolutions. Going back a few centuries, we would highlight that:

- **The First Industrial Revolution** (from around 1760 to around 1830) was marked by machines manufacturing technologies powered by steam or water that took over some manual methods. Mechanization boosted productivity by transforming industries, with a significant impact on people's lives.
- **The Second Industrial Revolution** took place roughly a century after the first one (from around 1830 to around 1960), with electrical energy being the leading technology. Electricity enabled the enhancement of the first industrial revolution innovations by leading towards mass production.
- **The Third Industrial Revolution** (from around 1970s to around 2010) is remembered as the automation era with electronics and information technologies being the leading ones. The Internet has played a pivotal role in transforming many social, industrial, commercial, government, and other sectors.
- **The Fourth Industrial Revolution** (from around 2014 to today) is widely recognized as the era of cyber-physical systems supported by data and analytics. It is still unfolding.

The selected technologies are part of a group marking our current period of the fourth industrial revolution or industry 4.0, even though some people are already talking about the emergence in the horizon of the fifth revolution or 5.0 characterized by people working seamlessly alongside robots and smart machines. But away from the labels, we will briefly review the selection as technologies of the 21st century. It is worth sharing some thought about the following technologies:

- Artificial Intelligence (AI)
- Internet of Things (IoT)
- Artificial Reality (ArtR™), which covers Augmented Reality (AR) Virtual Reality (VR), Mixed Reality (MR), Simulated Reality (SR), etc.

- Additive manufacturing or 3D Printing, and
- New Materials

14.1.1 Artificial Intelligence (AI)

In Latin, "intelligere" means to understand or to realize, and "intelligens" is about discerning. Both words point to the faculty or ability, generally reserved to humans, to learn (acquiring knowledge), reason, perceive relationships, grasp meanings or truths, etc. So, intelligence is the aptitude to understand and discern the inner side of what is being considered, whether it is an object, a topic, a situation, etc. As machines are programmed to process and act upon a large amount of data, the comparison to humans has taken place, and the notion of intelligent machines is now part of our vocabulary. However, a distinction has been made between artificial and natural intelligence. Keeping humans as a reference, artificial intelligence is about leveraging computation to model the brain processes and give machines some ability to act upon information provided to them to communicate, solve problems, etc. Even though machines may not think like us or behave exactly like us, they can be made to do things as we do in some situations. They can perform functions involving information processing such as voice recognition, sensing, vision, image processing, etc., for specific applications and, much more.

To achieve this, AI experts are pursuing several avenues. Since the technology is still evolving and will continue for decades to come, there are mostly two classifications of AI technologies depending on either their technical behavior as machines or their imitation and closeness to human beings:

- The first classification, summarized in Fig. 14.1A, recognizes Narrow AI applied to specialized fields to solve specific problems. Then, General AI reaches across several technical fields and synthesizes knowledge of various domains. Lastly, Super AI conceptually will reach a level far superior to human intelligence due to faster processing speed and much larger memory.
- The second classification, represented in Fig. 14.1B, covers stages that range from machines that react to stimuli without any capability to base their actions on previously accumulated data-driven knowledge to machines with self-awareness, emotions, and self-determination similar to a human-like race. Whether humans' creation can achieve

such an AI advanced state is to be determined in the future.

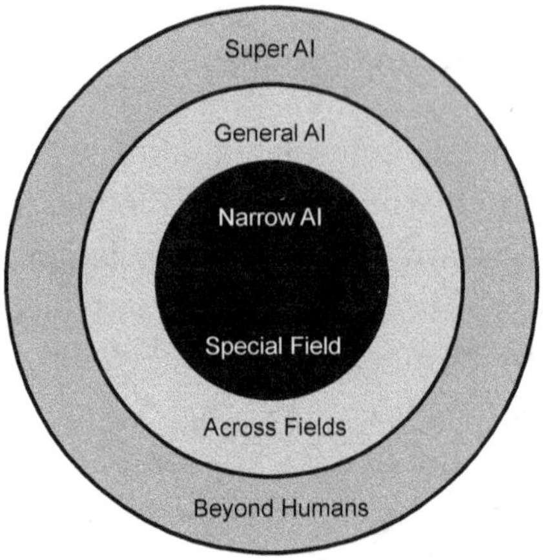

Fig. 14.1A AI Machine-based classification

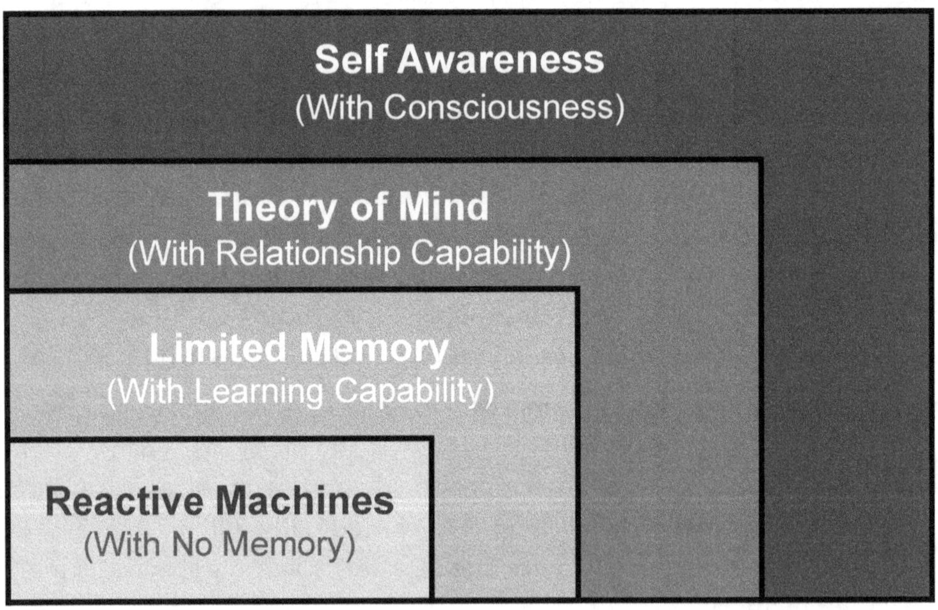

Fig. 14.1B AI Human likeness classification

Today, AI applications are all around us. We make calls and have machines answer them and guide us through a whole set of selections to direct our demands to the right people in the company or even answer them thoroughly to our satisfaction. Face recognition is another application addressed at many airports around the world and other sectors like banking. AI technologies will continue to evolve, and companies can no longer ignore them no matter what they sell.

14.1.2 Internet of Things (IoT)

The ubiquitous nature of the Internet is such that there is no need to dwell on its history. However, its application that has blown every prediction in the last decade and relevant to review in this book's context is its extension into data transfer over a network. It has created a worldwide network by connecting devices, one-to-one, one-to-many, and many-to-many. Such a capability made all physical entities able to exchange information and be recognized by their unique identifier, for example, a MAC address, an IP address, a Universally Unique Identifier (UUID), etc. Among the enabling technologies, we can mention:

- The miniaturization of electronics through new materials such as Silicon Carbide or Gallium Nitride or even nanometer-scale fabrication techniques.
- The perfection of radio-frequency identification (RFID) capable of identifying tags on objects.
- The creation of every kind of sensors of minuscule dimensions for temperature, light, motion, pressure, energy harvesting for powering sensors, etc.
- A whole suite of communication technologies and protocols such as Bluetooth, Wi-Fi, Zigbee, Cellular, LORA, Near Field Communication (NFC), etc.
- Improvements in embedded systems.

To all those technology wonders, we can add capabilities such as:

- Big data analytics with the availability of ever-improving algorithms,
- Cloud computing, which enables complex computations without having to attach large computers to the entities being analyzed.

With several billions of objects now connected, the new world enabled by IoT will only continue to expand through various innovations. From smart homes, self-driving vehicles, parcel or cargo tracking, smart cities, to smart crops, the list is endless. For example, smart homes can be managed from anywhere. Climate control, remote door opening and closing, security monitoring, remote sprinklers activation, and more add homes to the list of systems where telepresence is already a reality. The family of wearables continues to grow, and smart appliances are still in their infancy. Technologies only need the functions they can perform to drive innovation.

14.1.3 Artificial Reality (ArtR™)

The time has now come for each of us to pick our reality. When humanity's history is told in the future, the 21st century will be characterized as the era of alternative realities. People will choose from "Natural Reality" or from "Artificial Reality."

- Natural Reality (NR or NatR) was the only reality we knew. It is the reality that surrounds us, the one perceived by our natural senses and in which we live every day. It is our natural world in which we see, hear, smell, taste, and touch. We can hug a person and feel him or her. Take a minute to think. Those are all functions performed by biological sensors we are made of; they are part of our body and our being. Now, technology is evolving to perform some of the same functions with human-made sensors. These systems are no longer natural but artificial.
- Artificial Reality (ArtR) technologies are rapidly evolving to give us a sense of sight, hearing, touch, smell, and taste. Modern sensors provide us with visual, auditory, olfactory, haptic, and somatosensory (feeling pain, motion, temperature, etc.) artificially. They are now driving a whole new set of realities beyond the natural one we know. Today Virtual Reality (VR), Augmented Reality (AR), Mixed Reality (MR), etc. are being used in endless fields such as gaming, industrial, entertainment, education and medical. We are limited only by our imagination. Here again, functions and technologies are driving new realities.

14.1.4 Additive Manufacturing

Additive manufacturing, also known as 3D Printing, is not a new technology. In 1984, Bill Masters, an American inventor, filed a patent about automated manufacturing processes and systems based on computer support. The patent is so fascinating that part of its abstract, from the United States Patent and Trademark Office (USPTO), is quoted here: "A computer-automated manufacturing process and system (CAMPS) is disclosed which includes a computer system, which consists of a computer-aided design computer, and a machine controller, which receives a file of coordinate information. An article is designed by means of the computer-aided design subsystem, and a data file of three-dimensional coordinate information is compiled. The coordinate information is input to the machine controller, which controls servos, and, in a polar coordinate system. The servos further control the position of a working head (24) and working head (26) so that mass particles are injected to arrive at predetermined coordinate points in the coordinate system to form an article. An origination seed is fixed at an origin of the coordinate system, and the article is built up around the origination seed."[1]

This patent set the principles and foundation of 3D Printing. Other additive manufacturing inventions followed during the same period. The additive method worth mentioning and used the most until the recent explosions of the technology is stereolithography. It consists of stacking photopolymers layers and curing them with lasers. The next steps of the technology have evolved in different directions:

- Digital file format
- Printed materials
- Post-processing

ASTM (formerly, American Society for Testing and Materials) has summarized the structure of the standards including the materials, the processes, and the application domains. It is better to talk about Additive Manufacturing technologies in the plural rather than using a singular word based on the current state. As shown in Fig. 14.2 below, there are seven basic processes applicable to printing, either polymers or metals, for industrial applications. It is important to differentiate between industrial applications and other applications. The same principles that govern the processes mentioned are being leveraged to print different materials, including tissue (to print organs), food, houses, and the list could be extended.

Additive manufacturing has opened the door to endless applications. We have only scratched the surface. Complex shapes that conventional manufacturing processes could not address are now being realized because of

1 USPTO Patent Number, US 4665492A

additive manufacturing.

Once functions are defined, technologies can be applied to drive innovations.

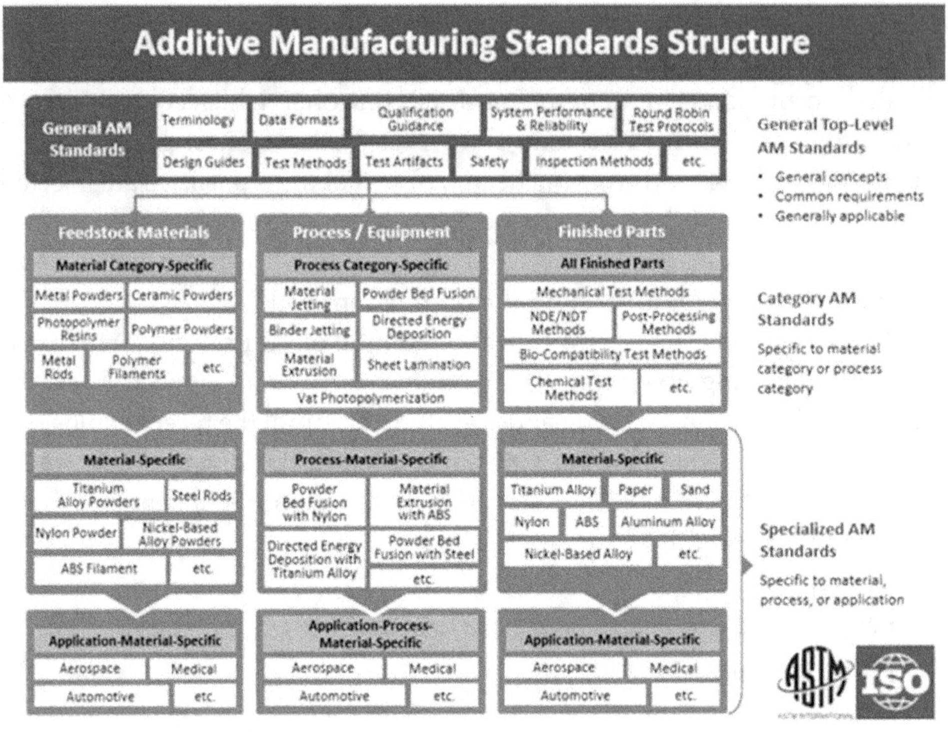

Fig. 14.2 Additive Manufacturing Standards Structure (Source www.astm.org)

14.1.5 New Materials

An exciting field not enough talked about is the new engineered materials created daily. Those who use the periodic table of chemical elements can appreciate that beyond the 94 elements called natural elements because they come from nature, new materials continue to be added after being engineered in laboratories. Chemistry successes highlight that humans continue to explore the new materials field and uncover materials with remarkable properties. Looking beyond the periodic table, some new materials with applications not too distant in the future can profoundly impact our lives:

- **Graphene**: even though graphite was invented in the 16th century and used by all of us as pencil lead, graphene is new. Discovered by Andre Geim and Konstantin Novoselov from the University of Manchester, the UK in 2004, they received the Nobel Prize in Physics in 2010. They realized that a tape they used to polish graphite had flakes of the material on it. Graphite is made of stacks of atom-thick layers of carbon, and as they removed additional graphite layers, they got to a material with only one carbon atom thick seen under an atomic microscope. The newly discovered 2-D material, which looked like a two-dimensional honeycomb, exhibited some remarkable properties that have scientists worldwide thinking of all the functions it could help perform and all the innovations their combination (functions and technology) could generate. Graphene is more robust than steel; it is flexible, transparent, and several folds better conductor of electricity than copper, and more. Graphene properties have accelerated research in various applications in solar cells, improved life batteries, filtration, electronics, coatings, touch screens, etc. Graphene is a material to watch for the transformative impact potential it could have in this century.

- **Metal-Organic Frameworks (MOF)**: as expressed by their name, Metal-Organic Frameworks are a whole class of engineered solid materials made of three-dimensional (3D) metal ions held by organic molecules. They are often referred to as organic-inorganic hybrid materials with essentially hollow and uniform structures that make them porous. Their porosity can be tuned, and their topology can be controlled during their design, giving them a large specific area compared to their mass and other unique properties. Published research shares their area to mass ratio as seven thousand square meters per gram of material. It means that a pound of MOF can be spread over 34 million square feet.

Several applications are being explored, from gas (such as carbon dioxide) capture, storage, separation, purification, adsorption to supercapacitors, sensing, lithium-Ion batteries, drug delivery, etc. Like present polymers (plastics), MOF's could generate thousands of variations in the future. As companies continue to research new ways to manufacture MOFs whiling bringing their cost down, they are already being defined as 21st Century materials. Researchers are still figuring out the many functions they could perform.

- **Carbon Nanotubes**: like graphene, a carbon nanotube (CNT) is made of a thin film of connected atoms of carbon but rolled into a cylinder. Therefore, it is a three-dimensional or 3D material. Some researchers define a CNT as a sheet of graphene rolled into a cylinder. Also, CNT's can be single-walled or multi-walled. Both materials have excellent properties and are being considered for different applications. CNT's are being explored, due to their strength and flexibility, for mixing into composite materials for applications such as clothing, super light, and strong ropes, concrete rebar, bulletproof vests, etc. They are also being explored for applications in sensors, battery electrodes, and transistors due to their excellent electrical properties.

In this discussion, carbon nanotubes are distinguished from graphene only in terms of their applications. Both are considered wonder materials, and their applications could dominate the 21st century.

- **Aerogels**: on the list of materials that could mark our century, aerogels as a class of materials are among the top ten. An aerogel is essentially a gel from which the liquid constituent has been removed and replaced with air or gas. The process generates a porous solid with an extremely low density. They can be made from many substances, and their production technologies continue to grow in numbers. Some silica, carbon, or iron oxide applications and other types of aerogels are already being used in many fields. What all of them have in common are:
    - Extremely low density as solid materials,
    - Extremely low thermal conductivity,
    - Extremely low speed of sound through any of them, etc.

Technology being the second element of *WinnoveX*™ equations, it was essential to describe a few selected developing technologies that have the potential to bring substantial changes to our everyday life.

In the next section, we will review a few innovations that display how the right combination of functions and technologies can generate unique solutions and help companies build a sustainable competitive advantage. Even if the functions existed for some time, choosing the appropriate technologies to perform them could allow companies to launch market winning solutions.

14.2 INNOVATION

Having explained *The WinnoveX™ Way* through its equations based on good execution of functions and their performing technologies, it would make sense to end the book with a section dedicated to some remarkable innovations beyond the previous chapters' examples. The list of innovation candidates was hard to finalize. There are so many systems that did not exist only a few years ago but are now part of our daily lives. We may be wondering how we could live without them. The criteria adopted was to select examples of innovation from various domains to highlight the breadth of *The WinnoveX™ Way* applications, some related to consumer products, industrial, and people. The choice landed on the innovations below, hoping that they will be representative of the cases that are made:

- Smartphones,
- Social media, and
- Robotics

14.2.1 Smartphones

When several people were asked to tell what smartphones have made obsolete in their lives, their reactions were immediate. Their input is summarized in Fig. 14.3.

The miniaturization of electronics and the digitization and storage of analog information (stored as prints, photos, and tapes, etc.) have enabled what could be considered one of the most important innovations of the 21st century so far. What is remarkable is how all the functions performed by various devices are now performed by only one device in a well-executed and integrated combination of hardware and software modules. There is a lot to learn in how smartphones transformed existing systems. Their example is not an isolated observation. To stay relevant and competitive more than ever before, companies in this century must pay attention to their usual competitors and those indirect competitors with technologies capable of performing customer value functions. The battle should not be limited to a finished product. It starts from the functions and the technologies that could perform them.

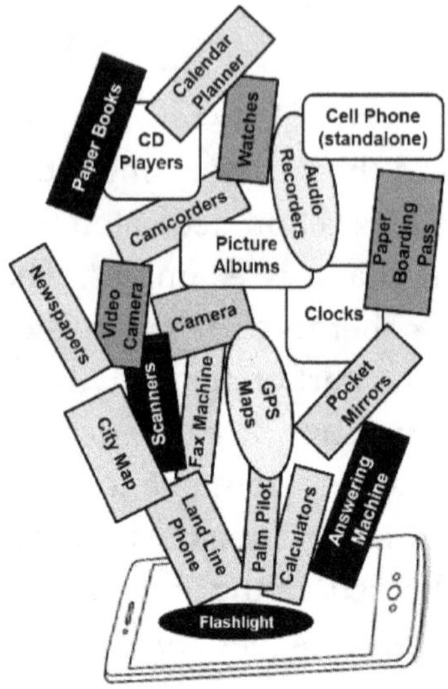

Fig. 14.3 List of some items made obsolete by smartphones

Fig. 14.4 is a way to represent the perfect application of the *WinnoveX*™ equations. The functions performed by many other devices for years are now handled by miniaturized hardware or by software modules.

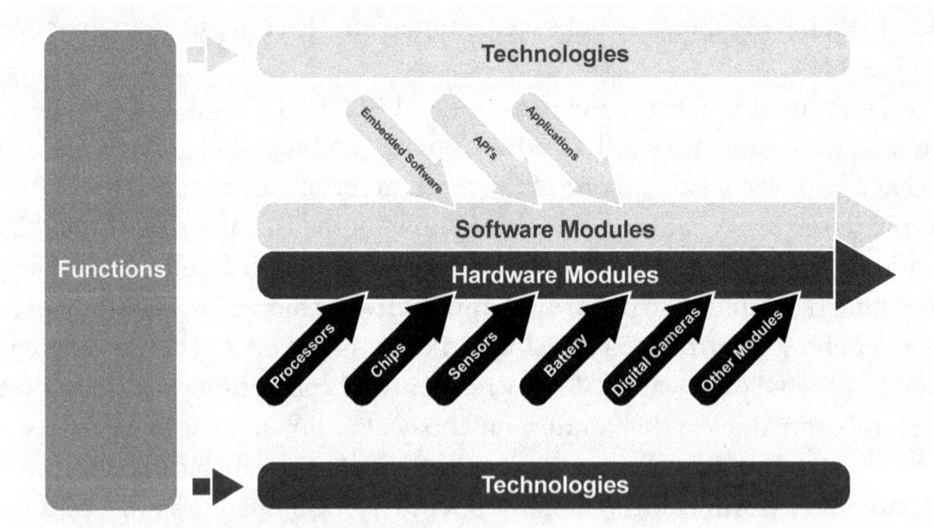

Fig. 14.4 A smartphone's functions and technologies

It is time for many companies whose products are mostly made of bulky electromechanical modules to question their legacy design and explore how the same functions could be performed through enhanced electronics and software. If they do not, someone else will. The difficulty is how to identify all the functions. Some functional analysis methods were presented in Part I of the book. Not every company appreciates functional architecture's strategic nature, yet it is at the heart of design and innovation. Companies should perform their products' functional analysis and determine their value-driving functions. Investing to understand the functions to commercialize is critical to the long-term competitiveness of any company. Then, technology investments cement its relevance to customers by keeping the products up to date.

Examples of such a practice are evident with handheld devices, cars, and other consumer products. Smartphone manufacturers release new versions yearly to create and sustain a high Emotional Resonance Index (See Section 2.2). The same thing is observed in the automotive industry with car manufacturers providing annual updates of a model on one or several brands. By collecting feedback from the users, they assess, every year, which technologies or designs to improve with existing functions. The addition of new functions is usually on a much longer cycle. It requires educating the customers through marketing to highlight the values offered. New functions provided by ABS, Airbags, self-braking, and other technologies were heavily advertised and marketed before the broader market reacted favorably beyond the early adopters.

14.2.2 Social Media

It is hard to imagine the world today without social media. As mentioned in Chapter 13, Aristotle, the Greek philosopher, characterized humans as social animals due to our need for gathering, communicating, and organizing our lives as groups. Our need to gather and communicate has found its response in modern technologies that enable virtual gathering without the limitation of being in one location.

From a purely functional viewpoint, the media was about means to "Share Information." It was not about one "medium" but several "media" to inform people. In our new social era, it has assumed the function to "Connect People" when seen as social media, emphasizing social interactions.

Therefore, the main purpose function of social media is to "Connect People", which can then be decomposed into:

- "Share Information" (text or voice messages, pictures, video, blogs, etc.). It can be said that information sharing has become just one of the functions when social interactions take a more central place with social media.
- "Network People" (social groups, communities, organizations, etc.)
- "Transact Business" (selling or buying products or services, advertising, etc.)
- "Educate People" (online courses, live classes, online exams, etc.). We could argue that educating people is sharing information. However, the information provider expects the recipient to acquire the knowledge and do something with it for a greater good. Educators have expectations that generic information providers do not have. Newspapers, radio, or television companies are not explicit about any obligation on those who receive their news to act upon the information they provide.
- "Examine Patients" (telemedicine, online files, etc.), etc.

Each of the functions can be decomposed further, but the examples of applications shared show the ramifications of social media purpose function to "Connect People."

These functions are performed by a whole suite of technologies that have today created one of our largest industries. Technologies provide a whole body to functions they perform to satisfy specific needs. Like smartphones in the previous section, social media continues to grow and invade other fields. The means to share information through social interactions may evolve even though the functions they perform may be the same.

14.2.3 Robotics

Humans' fascination for robots transcends every culture and geography. So, considering innovation examples to illustrate the *WinnoveX*™ theory, it was natural to include "Robots." The concept of machines that would serve people and help them solve their challenging problems has captured people's imagination for decades. Isaac Asimov defined the three laws of robotics.[2]

---

2 Isaac Asimov, *Runaround*, included in the collection I, Robot 1950

He suggested that:

1. A robot may not injure a human being or, through inaction, allow a human being to come to harm.
2. A robot must obey the orders given to it by human beings except where such orders would conflict with the first law.
3. A robot must protect its own existence if such protection does not conflict with the first or second law.

Asimov intended to set, in a fictional world, the basis for the relationship between intelligent machines and humans. Even though other people suggested various laws, the three quoted remain my favorites. Over the years, my own definition of a robot has evolved. It can be stated as: An intelligent machine with the capability to interact with humans to receive input and autonomously execute complete missions.

In the context of *The WinnoveX™ Way*, robots' functional architecture can be complex. However, considering the intent behind their creation throughout history, one can say that their main purpose function is to "Augment Human Capabilities." By putting humans at the center of robots' development, a few observations should be made:

- Humans perform countless functions, and if robots are designed and programmed to augment human capabilities, they should be adapted to many purpose functions. Whatever humans do, robots should be conceived to do better and faster in many fields of activities. They can be imagined to be everywhere.
- Robots should be directed to serve humans and, therefore, be built from technologies that preserve human safety in line with Asimov's first law.
- The mission any robot is conceived for should dictate its architecture, just like motor vehicles' architecture is different from machine tools or any other industrial equipment's architecture.

Whichever way the functions of a robot are identified, grouped, and categorized to define their modules, some subsystems are to be considered for every one of them are as follows:

1. Human Interaction Subsystem
2. Intelligent Controls Subsystem

3. Propulsion Subsystem
4. Structural Skeleton Subsystem, (even if they are not humanoids, they still need a frame or structure, called here the skeleton)
5. Object Handling Subsystem
6. Power Management Subsystem, (functions performed to manage the machine's autonomous power if not plugged into a fixed source)
7. Safety Subsystem
8. Health Management Subsystem

The three ingredients, functions, technologies, and execution, always converge to deliver winning innovations, as seen through all the examples. And that is what *WinnoveX*™ is all about.

14.3 21ST CENTURY TECHNOLOGIES AND INNOVATIONS SUMMARY

Having presented the way to *Winning Innovation Excellence* or *WinnoveX,*™ it was inevitable to dedicate a chapter to a few technologies and review some innovations to show how functions are performed to make them happen. Technologies that may have existed in some fashion for years are seeing outstanding growth with unprecedented applications. A few were selected and briefly reviewed, mainly:

- Artificial Intelligence (AI),
- Internet of Things (IoT),
- Artificial Reality (ArtR™) that covers Augmented Reality (AR), Virtual Reality (VR), Mixed Reality (MR), etc.
- Additive manufacturing or 3D Printing, and
- New Materials

Beyond the exponential growth of these technologies, the 21st century is also seeing their convergence from various domains generating solutions that continue to augment human capabilities.

The technologies were briefly described, followed by three innovation examples from various fields to share how *WinnoveX*™ equations universally apply across domains and industries. The choice landed on:

- Smartphones,
- Social media, and
- Robotics

As each of the innovations was reviewed, it emerged that the functions performed by any entity should be identified, and its architecture should be part of a company's assets. Such an investment is necessary for a company to stay current, relevant, and competitive. Fig. 14.3 shows a few devices that were absorbed and made obsolete by smartphones. It is a reminder that "indirect" competitors are as much a threat as the known and direct ones. They could simply be silent killers of once-great companies.

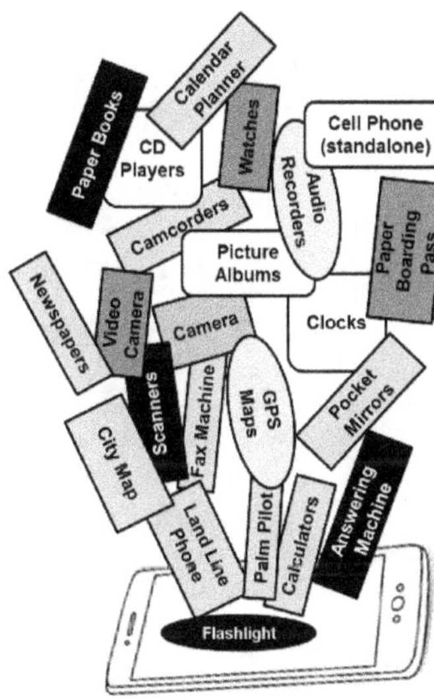

Fig. 14.3 List of some items made obsolete by smartphones

The battle for relevance and competitiveness is not limited to finished products, but it starts from the investment in functional analysis, the right technology choices, and rigorous execution. Companies are encouraged to explore all their "indirect" competitors who can perform their value-driving functions from another industry. Failure to do so can easily make them road kills.

15

Conclusion

Humans have continuously improved their conditions of life through innovation. History has uncovered artifacts that showed how every society and every generation of people had accomplished incredible innovations by finding solutions to their unmet needs. The primary motivation of the project that generated this book was about understanding the principles at the core of any innovation. Such an understanding would set "an Innovation Blueprint" and make it systematic instead of accidental or simply left to a few people's genius or creativity.

The approach taken started from observations. Observations are to be practiced intentionally and regularly. As stated in the Forewords, I do not spend a day without dedicating at least 60 minutes to observe things around me. They could be anything from nature (flowers, animals, trees, people in their interactions, clouds formations, etc.) to human-made systems (cars, traffic, airplanes at airports, clothes, houses, etc.). It is nice to watch a movie in a theater, but it is gratifying to enjoy the natural movies happening around us in nature's amphitheater. The daily observations make me present and conscious of the world around me. I encourage everyone to do so. These observations reinforced my learning from formal education that everything any entity (natural or artificial) does is through performing functions. The difficulty is in identifying the functions. Chapters 2 through 4 are dedicated to several techniques to identify functions performed by entities. Understanding any being's functions opens its world to the trained eyes and minds. People who have developed the skills can analyze existing systems and understand their functions, but more importantly, they would also be able to take the main purpose function of any system's and decompose it top down to the lower level functions whose fulfillment would enable the top-level one. Such ability is critical, as shown along the book, to fully represent how existing things work, or what new ones should do. Once the functions of a system are identified, they are grouped into independent sets that are then embodied by physical entities to perform them. The entities that perform the independent function sets are any system modules and represent its building blocks. The modules that work together to perform a higher-level function form a subsystem, and the subsystems, in turn, form the system. Modules have dual nature. They are both functional as well as physical entities. They are functional because they are created by the functions performed and physical since they can consist of hardware and software. Modules and their relationships or interactions form the architecture of a system. Chapter 3 shows examples of architectures and their representations. The functional architectures of any system represent one of the main elements of its DNA. Once they are understood, the relationships between modules can be changed to create different embodiments, generating innovative solutions. A modular architecture is one of a system based on modules. It has numerous advantages:

- It enables the design and offering of many product configurations from a limited set of modules, which is key to fulfilling mass customization.
- A modular product architecture puts modules development and systems integration on two separate channels, allowing systems integration teams to shorten a product's time to market.
- The application of modeling and simulation to modular systems enables the creation of virtual labs to digitally test a full system, its subsystems, and modules. Such a capability cuts development cost and allows design teams to try many solutions at lower risks. Virtual laboratories are crucial in a system's life to enable the virtual testing of almost all its instances and their variants and over time after any change.
- Once the functional architecture of a system is defined, it applies to its manufacturing, supply chain, logistics, product management, and service. Chapters 7 through 11 explored and described the opportunities presented by such applications.

Winning innovations happen through the successful execution of functions performed by appropriate and selected technologies. But before executing functions and technologies, the unmet needs must be understood. Then, they must be expressed through the functions that need to be performed to satisfy them. When the unmet needs are mapped to the type of solutions companies can choose to meet them, six categories of innovation emerged, as shown in Fig. 6.4. Incremental and radical solutions call each for different approaches to address them. Two categories that are worth mentioning are both characterized as unknown, unmet needs incomprehensible to the customers. The first one, Category 5, is addressed with incremental solutions, and the second, Category 6, is addressed with radical solutions.

Companies should tackle category 5 with caution. It attempts to provide an incremental solution to an unknown, unmet need that is also incomprehensible to customers even when expressed to them. As presented in Chapter 6, the return may not be worth the investment. Category 6 comes with big rewards but equally with some risks. It would require investments to educate the customers and users of the future system. The return is worth the effort, but execution must be well resourced. A failure would allow fast followers to harvest the fruits of any investment by building upon any progress made.

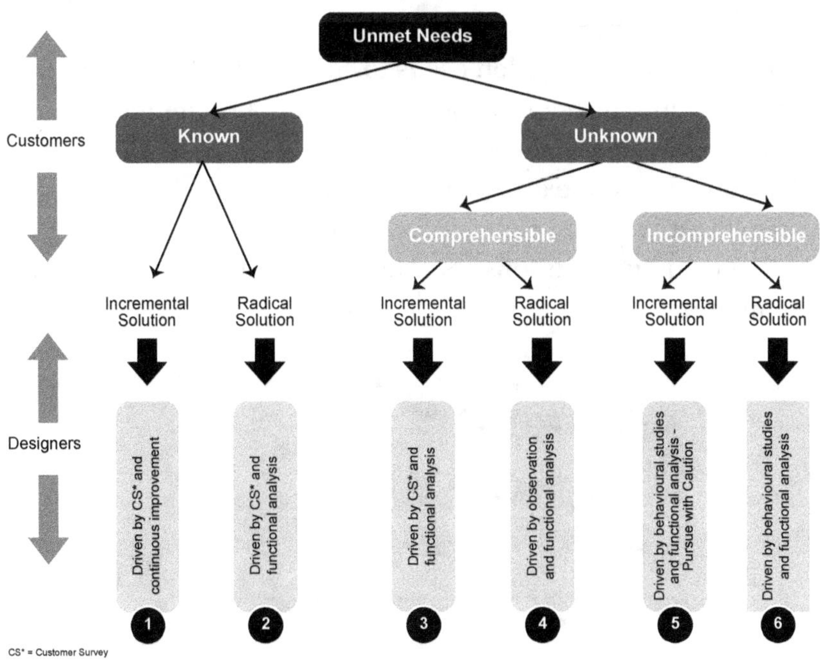

Fig. 6.4 Unmet needs categories and solution types with their drivers

Once the investment is made, and the functional architecture of a company's systems is defined, the fundamental equations of innovation can be fully leveraged. As defined in chapter 6,

(1) Functions + Technology = Innovation

Equation (2) is generated by taking execution into account. We have:

(2) (Functions + Technology) x Execution = Winning Innovation

Equations 1 and 2 are defined as winning innovation or *WinnoveX*™ equations, and they are at the core of *The WinnoveX*™ *Way*. When the functions of any system, existing or new, are mapped to selected technologies and well-executed, innovations are sprouted, and the process is sustainable.

All three key elements must be present:

- **Functions**: their identification requires training and in-depth knowledge of the system of interest. It must be done by the subject matter experts and as a team exercise.
- **Technologies**: their selection needs to be preceded by a whole landscaping exercise in which all the potential candidates are reviewed against the functional requirements.
- **Execution**: a good execution combines the discipline of program management with the rigor of technology and product development to deliver unique solutions that delight the customers and energizes the employees.

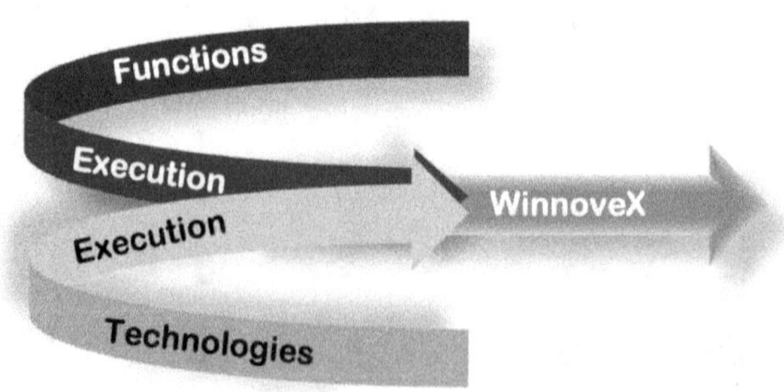

Fig. 6.5 The WinnoveX™ Way

Anyone of the three elements without the other two would not yield optimum results. Therefore, the *WinnoveX™ Way* is based on all three components. The process is defined in Fig. 6.7; a journey that starts with identifying the functions performed by existing systems (products, processes, or service delivery), or their definition for new systems, through functional analysis. Once the functions are defined, and the functional architecture is set, requirements can be determined. Functional requirements are the functions and the metrics that express how well they should be performed to deliver a competitive or market-leading system. With the functional requirements, technology landscaping is conducted to review all the technologies with the potential to perform the functions in their best embodiments. The exercise can be conducted at a module, a subsystem, or even at a system level.

The best technologies are then selected and mapped to the functions. It could be just one function, as well as a set of functions.

Value engineering experts set a direct correlation between function and value. Therefore, determining a system's functions also leads to understanding the value being offered to customers.

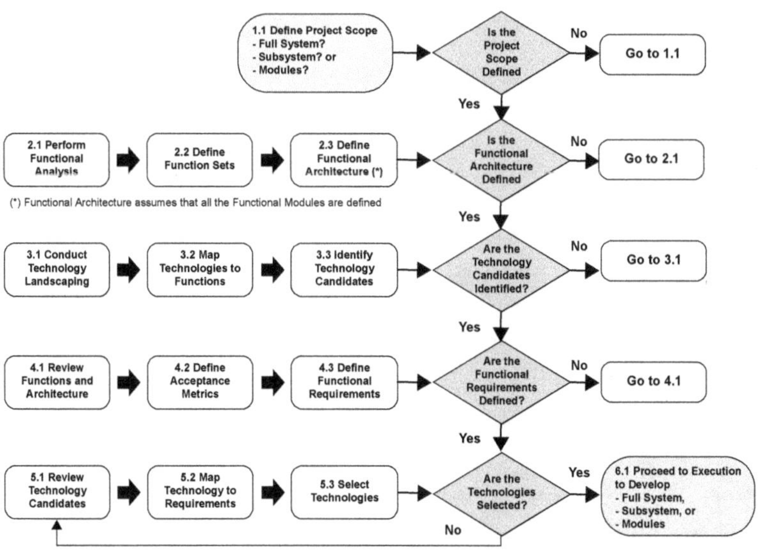

Fig. 6.7 The WinnoveX™ Way Process

System development should start after the functional requirements are defined, and the right technologies mapped to them. An essential enabling asset of the *WinnoveX™ Way* is a Modular Product Architecture (MPA). With such an asset, the field of innovation should be defined at the appropriate level (module, subsystem, or system).

The concepts developed and presented in the book highlight three opportunities with the power to give a competitive edge to companies that invest in embracing them:

- Winning Innovation Excellence through *The WinnoveX™ Way,*
- Mass customization through a modular product architecture and,
- Creation of brand loyalty through a high Emotional Resonance Index (ERI)

Appendix

CHAPTER 2: SYSTEMS ANALYSIS

A.1 Addendum to Section 2.1.2 – Functions Identification and Decomposition

A1.1 Operational Scenarios Review (OSR) Method – The elevator example

Full Table 2.1 of an elevator user scenario for a single trip normal run from a group of units.

Another way to identify a system's functions is through the breakdown of existing products.

Step	User Visible Behavior	System Behavior	Functions
1	Passenger presses Hall button		Request service
2		System recognizes service request	Acquire request
3	Hall button "tell-tale" light activates		Acknowledge request
		System assigns a car to service request	Dispatch elevator
		System opens the brakes	Move car
		Car travels to service request location	Move car
4	Hall light activates	System activates hall light	Indicate arrival
5	Car door open	Car stops at service request location	Stop car
		Drop the brakes	Stop car
		System opens doors	Allow access
6	Passenger enters car	Keep doors open	Allow access
7	Passenger presses button in car		Select destination
8	Car button light activates	System recognizes destination selection	Acknowledge request
9	Car and hall doors close	Secure passenger in the car	Protect load
10		System opens the brakes	Move car
11	Car moves	Car travels to destination	Move Car
12	In car information displayed	Display direction and position information	Inform passenger
13	Car stops	Drop the brakes	Stop car
14	Open doors	Car and hall doors open at destination	Allow egress

Table 2.1 Elevator user scenario for a single trip normal run

A1.2 Existing Products Breakdown Analysis (EPBA) Method

A1.2.1 Detailed steps description

The details of each step are described below:

- **Step 1**: Decompose the existing products into their building blocks

The method starts from the physical embodiments of the products that are then broken down into their legacy elements. The elements could have been derived from how the product was designed and assembled over time at the sourcing factories or even from their assembly processes. The legacy breakdown serves as a reference for future functional analysis. An existing bill of materials becomes particularly useful for the exercise.

- **Step 2**: Determine the functions performed by each of the blocks

In general, the product offerings with the same or similar main purpose function or operating mission function share many lower-level functions. Therefore, the analysis is done on the grouping of all the systems that share similar operating mission functions and are offered as part of the company's product portfolio. The groupings may also be made around the same governing principle or law of physics or leading technology. The various elements are decomposed from the high-level offerings to lower and lower-level ones. Every element's functions are then determined. The exercise allows the design teams to fully grasp what their systems do. It also highlights what lower-level elements do to enable the actions of a higher-level one.

At the end of the exercise, many functions are not necessarily in the right order.

- **Step 3**: Regroup the functions from the lowest level parts without the constraint of how they were performed by the existing products, to a level where they can now be performed by a variety of technologies, to create the functional architecture

By extracting functions from the lowest level parts, one acquires functions that exist because of the previous design and implementation decisions. The existing functionality needs to be brought to a level 'above' where technology / allocation decisions were locked in to innovate with different technologies and implementations.

After grouping the functions, each set is given a name that generically represents the element that performs them in the system. It could be a module or a group of modules into a subsystem, or even elements making up modules. The new generic names can be the same or different from how they have been characterized in the company for years, depending on whether they represent the group of functions being performed. The exercise highlights the dependencies between the elements. The grouping of the functions is done based on coupling avoidance to assure the independence of the physical entities that will perform them. Such property enables the design team to test them in isolation from the whole system.

A1.2.2 Details of the EPBA method applied to a portfolio of bicycles

Physical Domain		Functional Domain	
Product Level 1	**Original Physical Element (Design Solution)**	**Functions**	**Lower Level Functions**
Frameset		**Carry Laod, Transfer load,**	
	Frame		Distribute load, Hold wheel,
	Fork		Transfer load, Hold front wheel, Absorb shock
Seat		**Enable seating**	
	Saddle		Provide seat, Carry load
	Seatpost		Transfer load, Hold seat, Enable seat adjustment
Luggage carrier		**Carry load**	
	Carrier		Carry load
Front wheel		**Enable motion**	
	Front wheel		Enable motion, Provide braking surface
Transmission		**Transmit motion**	
	Pedals		Generate torque, Start motion, Sustain motion, Rest feet, Propel bike, Reduce speed
	Chain		Provide power, Transmit motion, Vary speed
	Cassette		Select speed, Change speed
	Crankset		Convert motion, Transmit torque, Change speed
	Front Derailleur		Lift chain, Change gear
	Rear Deraiileur		Lift chain, Change gear
	Shifter		Select speed, operate derailleur
	Bottom bracket		Connect crankset, Enable crankset rotation
Drive		**Convert power**	
	Power Converter		Convert power
Rear wheel		**Enable motion**	
	Rear Wheel		Enable motion, Provide braking surface
Fender		**Deflect road water**	
	Fender		Protect cyclist (from water splash)

Table 2.3A Bicycle breakdown with the primary and secondary functions

Physical Domain		Functional Domain	
Product Level 1	**Original Physical Element (Design Solution)**	**Functions**	**Lower Level Functions**
Frameset		**Carry Laod, Transfer load,**	
	Frame		Distribute load, Hold wheel,
	Fork		Transfer load, Hold front wheel, Absorb shock
Seat		**Enable seating**	
	Saddle		Provide seat, Carry load
	Seatpost		Transfer load, Hold seat, Enable seat adjustment
Luggage carrier		**Carry load**	
	Carrier		Carry load
Front wheel		**Enable motion**	
	Front wheel		Enable motion, Provide braking surface
Transmission		**Transmit motion**	
	Pedals		Generate torque, Start motion, Sustain motion, Rest feet, Propel bike, Reduce speed
	Chain		Provide power, Transmit motion, Vary speed
	Cassette		Select speed, Change speed
	Crankset		Convert motion, Transmit torque, Change speed
	Front Derailleur		Lift chain, Change gear
	Rear Deraiileur		Lift chain, Change gear
	Shifter		Select speed, operate derailleur
	Bottom bracket		Connect crankset, Enable crankset rotation
Drive		**Convert power**	
	Power Converter		Convert power
Rear wheel		**Enable motion**	
	Rear Wheel		Enable motion, Provide braking surface
Fender		**Deflect road water**	
	Fender		Protect cyclist (from water splash)

Table 2.3B Bicycle breakdown with the primary and secondary functions

A1.2.3 Table 2.5 Bike functions, modules, and subsystems

The table summarizes the transition from the bicycles' physical elements to their functions, modules, and subsystems.

We can now consider the functional building blocks of the portfolio as the subsystems and modules, as shown in Table 2.5

Physical Domain		Functional Domain		Modules Functional and Physical	Subsystems
Product Level 1	**Original Physical Element (Design Solution)**	**Functions**	**Lower Level Functions**	**Functions to Physical Mapping into Modules**	**Modules Grouping to Subsystems**
Frameset		**Carry Laod, Transfer load,**			**Frameset**
	Frame		Distribute load, Hold wheel,	Frame	
	Fork		Transfer load, Hold front wheel, Absorb shock	Fork	
Seat		**Enable seating**			**Load Carrier**
	Saddle		Provide seat, Carry load	Seat	
	Seatpost		Transfer load, Hold seat, Enable seat adjustment	Seatpost	
Luggage carrier		**Carry load**			**Load Carrier**
	Carrier		Carry load	Luggage Carrier	
Front wheel		**Enable motion**			**Wheels**
	Front wheel		Enable motion, Provide braking surface	Front wheel	
Transmission		**Transmit motion**			**Propulsion**
	Pedals		Generate torque, Start motion, Sustain motion, Rest feet, Propel bike, Reduce speed	Pedals	
	Chain		Provide power, Transmit motion, Vary speed	Motion transmission	
	Cassette		Select speed, Change speed	Speed selector	
	Crankset		Convert motion, Transmit torque, Change speed	Motion converter	
	Front Derailleur		Lift chain, Change gear	Speed selector	
	Rear Derailleur		Lift chain, Change gear	Speed selector	
	Shifter		Select speed, operate derailleur	Shifter	
	Bottom bracket		Connect crankset, Enable crankset rotation		

Table 2.4A Bike functions, modules, and subsystems

Drive		**Convert power**			**Propulsion**
	Power Converter		Convert power	Power converter	
Rear wheel		**Enable motion**			**Wheels**
	Rear Wheel		Enable motion, Provide braking surface	Rear wheel	
Fender		**Deflect road water**			**Frameset**
	Fender		Protect cyclist (from water splash)	Fender	
Steering Assembly		**Control bike steering**			**Bike Control**
	Handlebars		Steer bike, maintain direction,	Steering assembly	
	Headset		Ease front fork rotation,	-	
	Stem		Attach handlebars,	-	
	Grip		Prevent hand slip	-	
	Acceleration lever		Vary speed	Acceleration lever	
Dashboard		**Human Machine Interface**			**Dashboard**
	Speedometer		Show speed	Speedometer	
	Charge indicator		Show state of charge	Charge indicator	
Brakes		**Stop bike**			**Brakes Assembly**
	Front brake		Reduce speed, stop bike	Front brake	
	Rear brake		Reduce speed, stop bike	Rear brake	
	Brake cable		Activate brakes	Brake cable	
	Brake levers		Apply brakes	Brake levers	
Generator		**Convert energy**			**Power Generation**
	Generator		Generate power	Generator	
Lights		**Illuminate road**			**Lights**
	Headlight		Illuminate ground	Headlight	
	Rear light		Signal presence	Rear light	
Engine		**Convert energy**			**Power Generation**
	Engine		Provide power, Move bike	Engine	
Tank		**Hold fuel**		Tank	**Power Generation**
	Container		Contain fuel		
	cap		Prevent fuel spill		
	Fuel		Operate engine	Fuel	
	Gauge		show fuel level		
Stopper		**Hold bike (at parking)**			**Controls**
	Stopper		Enable parking	Stopper	
Electric motor		**Convert Energy**		Electric motor	**Power Generation**
	Electric motor		Provide power, Move bike		
Battery pack		**Store Energy, Power bike**			**Energy Storage**
	Battery		Store energy, Power bike	Battery pack	
	Battery Charger		Transfer power	Battery charger	
	State of charge indicator		Show state of charge		
	Charge supervisor (SW)		Supervise charging	Charge supervisor (SW)	

Table 2.4B Bike functions, modules, and subsystems

A1.2.4 Alternative representation of the bike portfolio modules and subsystems

Fig. 2.7 is an alternative representation of Fig. 2.6 that allows the distinction of each subsystem with its modules.

Both representations are valid and are used based on the systems analyst's preference.

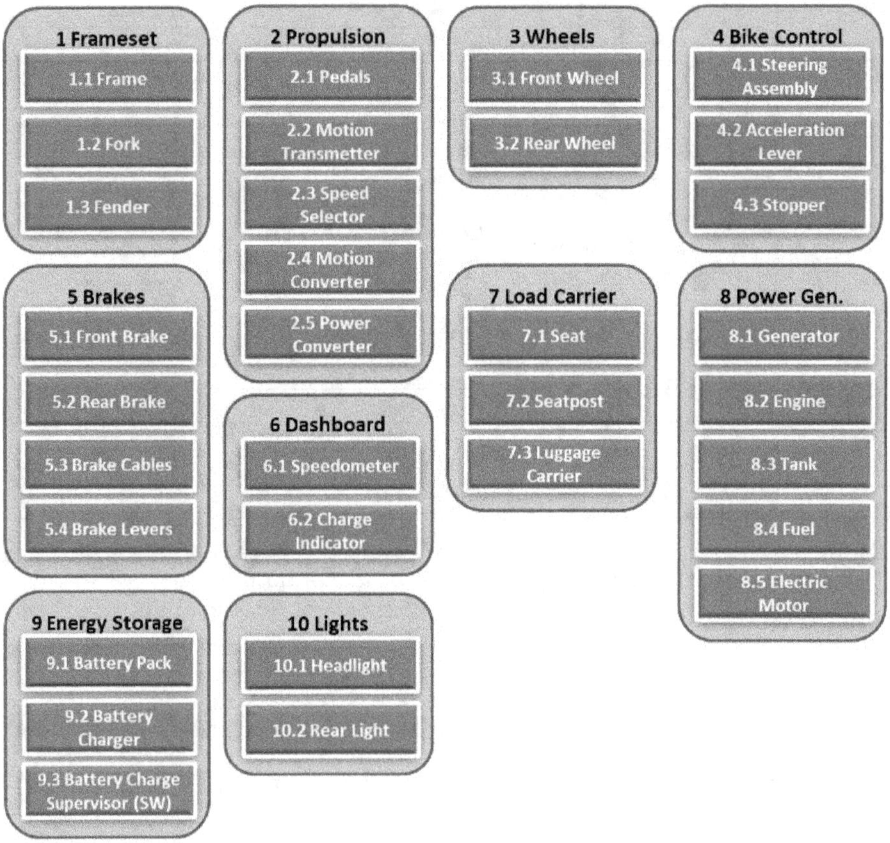

Fig. 2.7 Bicycle Subsystems and Functional Modules

A.1.3 Details of the Operand Flow Analysis Method

Another way to implement the Existing Products Breakdown Analysis is through the Operand Flow Analysis or OFA method. It is a subset of the EPBA Methods and a way of decomposing existing products into elements that lead to the definition of subsystems and modules.

The method assumes that by following a system's element that interacts with several of its constituents, the interactions' enabling functions can be identified.

Example: Vapor Compression Based Architecture (VCBA) of a Heating, Ventilation, Air Conditioning and Refrigeration (HVACR) system

Using the Operand Flow Analysis (OFA) method, most of the operands are:

- Refrigerant
- Power
- Controls
- Air
- Water (in the case of a chiller)
- Gas (for a furnace, etc.

All of which interact with multiple constituents of thc HVACR system.

Some steps of the process are omitted for simplification, but the results are exposed.

Refrigerant Flow

Whether used in buildings or transport refrigeration, most vapor-compression air conditioning systems share the refrigerant cycle illustrated in Fig. 2.8..

Fig. 2.8 illustrates the cycle that governs most air conditioning systems based on vapor compression. In this case, it is the thermodynamic cycle that is the unifying source of the systems. All of them have the same or similar operating mission function, which is to: "Condition Environment." The environment could be any living space (a building, a train wagon, a bus, etc.) or a box such as a marine container or even a refrigerated cell. The verb of the function "to condition" can mean "to heat", "to cool", "to ventilate", or to refrigerate. In this case, by following the operand, some critical elements that drive the system to perform the main purpose function can be defined.

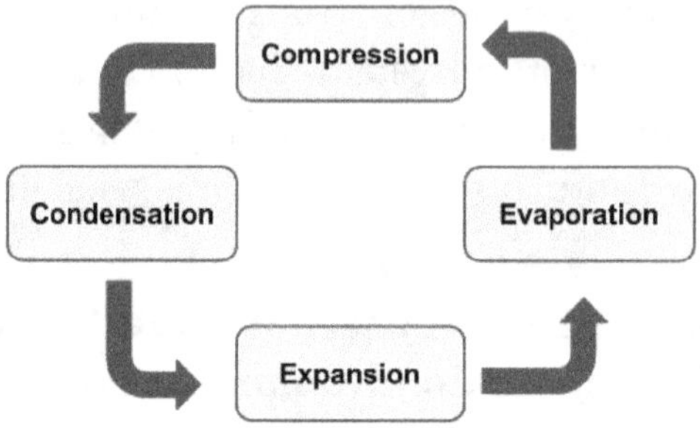

Fig. 2.8 Air Conditioning Refrigerant Cycle

- The refrigerant needs to be compressed. The function "compress refrigerant" is performed by compression elements. In this case, the function's highest-level action is "compression" done by a group of elements that make up a compressor generically independent of any technology. It does not matter whether it is a screw, centrifugal, scroll, or other technology. It is important to note that we are still at a functional level. We can pencil down that compression is a significant step toward the operating mission. Therefore, a compression subsystem will be functionally defined. As we will see later in the physical domain, several elements perform various functions that lead to "Compression." Therefore, we will also have a compression subsystem in the physical domain made of hardware and software.
- The refrigerant needs to be condensed. "Condense refrigerant" is the function or action that removes the heat the refrigerant carries after compression. It brings it to a liquid level. That function is vital because, without it, the next step would not happen as desired. Just like in the case of compression, the condensation subsystem is defined regardless of the technology.
- The refrigerant will then be allowed to expand. Therefore, the expansion subsystem is defined. As it expands, the temperature falls.
- The refrigerant needs to evaporate. It passes off in vapor changing from liquid to gas again. Several elements contribute to the "evaporate refrigerant" function. We will group them into the evaporation subsystem

just like we did for the other subsystems in the physical domain.

Just by following the refrigerant flow, four subsystems have been defined.

- Compression Subsystem
- Condensation Subsystem
- Expansion Subsystem and
- Evaporation Subsystem

It is important to note that, at no point, the process is about HOW the functions are performed. That is the domain of technology and is widely open to endless innovation. The companies that will perform those functions reliably at a competitive cost will win in the marketplace.

The same exercise with the refrigerant can be pursued with the other operands.

Power Flow

Electrical power brings energy to multiple elements of the system, and its distribution is critical for the actuation and performance of the system. "Power System" or "Energize System" can be further decomposed. Some lower-level functions are:

- Generate power (not part of the system if power is supplied from utility companies)
- Transmit power
- Convert power
- Condition power
- Distribute power
- Recover Power (Energy in general)
- Store Power (Energy in general), etc.

All those functions will be further decomposed. They will be used to define generic elements that perform them.

Without getting to design details, several subsystems could be defined:

- Powertrain Subsystem
- Power Distribution Subsystem
- Energy Recovery Subsystem
- Energy Storage Subsystem

As new elements are added to the structure, the functional architecture takes form. It is to be noted that Power Generation Subsystem is not considered part of the system since it is assumed that the electric grid will supply power. However, if it is determined that the system will work off-grid and be delivered with a generator of any type, such a subsystem should be considered.

Signal / Information Flow

For the system to perform as expected, its elements' logical intervention sequence must be well defined, either for a single unit or a group. It should also consider how operators should interact with the system to perform its main purpose function. The functions of this operand can be common to various systems across industries. They can be imported from other systems. There are two essential items to consider while analyzing this flow. How operators need to interact with the system and how the system's elements shall communicate with one another in a single system and a system of systems. At the end of the exercise, some of the generic elements defined could be:

- Control input
- Control output
- Compute algorithm
- Coordinate logic
- Communicate logic
- Collect data, etc.

New subsystems can be defined as:

- Human Machine Interface Subsystem
- Controls Subsystem
- Health Management Subsystem

Water Flow

In the case of a chiller, water is a crucial operand. It carries the desired temperature to the established locations and exchanges it with the ambient by absorbing its heat. It can be either cool, cold, or hot, depending on its location in the system. Following its cycle would lead to elements that other operands may not have detected.

It also happens very often that operands cross paths in their cycle.

An understanding of the functions performed by a system is revisited when the EPBA method is being used. Some engineers who have carried out designs from legacy all the sudden get their epiphany when they realize that some functions are misinterpreted during the exercise, and some opportunities emerge to improve a design a company held for decades. However, to get to the functions, a level of abstraction needs to be exercised to understand the system's elements.

The subsystem derived from the water flow can simply be a Water Cycle Subsystem or Water Flow Subsystem. We could be more creative in the naming, but the idea is clear. Some of the functions are:

- Contain water
- Move water
- Conduct water
- Cool Water

Here, essential considerations need to be made. Suppose the Water Cycle Subsystem becomes too large, like in the case of a large chiller where a cooling tower is required. In that case, the subsystem becomes a system by itself, and the product that ultimately delivers the main purpose function to the customer becomes a system of systems. Such an approach is important to avoid combining too large subsystems with ridiculously small ones to define a system. Such a distinction becomes useful for managing the system in terms of the supply chain, cost, installation, etc. as explained in Part II.

Gas Flow

Some units burn gas in their furnace to produce the heat necessary to bring an environment to the desired temperature. To complete the system's picture, we will define it without getting into the functions' decomposition.

Heat Generation Subsystem

When all the operand flows are analyzed, and all elements are reviewed, the subsystems' picture could be summarized in Fig. 2.9 as a Vapor Compression Based Architecture (VCBA) Main Subsystems.

Other subsystems have been added. Those are for practical purposes as

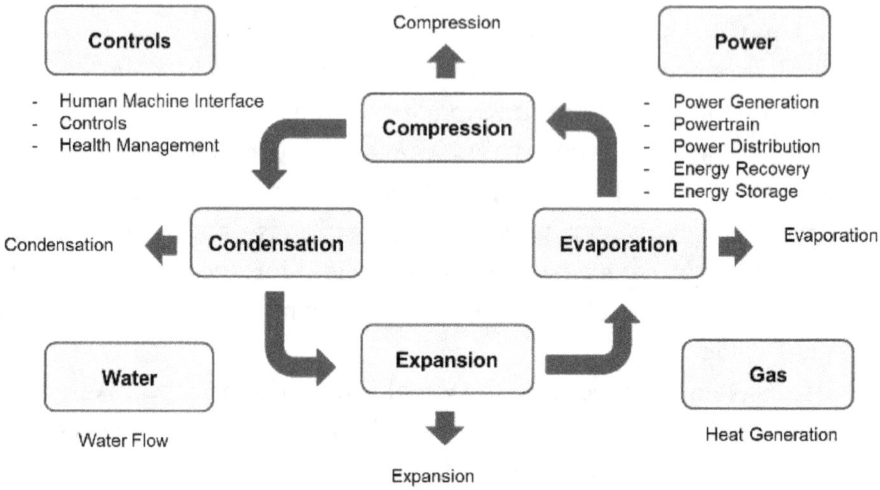

Fig. 2.9 Vapor Compression Based Architecture (VCBA) Main Subsystems

they are common to many systems across industries. They are not derived from the operand flows but other significant functions such as:

- Protect System
- Handle System
- Instruct Operator
- Etc.

They drive subsystems such as:

- Enclosure Subsystem
- Package Subsystem
- Documentation Subsystem

Packaging and documentation of a product should never be afterthoughts. They perform critical primary functions that are then decomposed to create necessary modules for the system's well-being and management. The packaging subsystem's design is critical to protect the system during loading and transportation. Inadequate protection of a system can compromise its integrity and its performance. The packaging subsystem is treated just like any other subsystem by understanding its functions and defining its functional and non-functional requirements. Likewise, the Documentation

Subsystem also needs to be carefully studied and designed to help customers, users, and all the other stakeholders make fair use of the product and operate it at its peak performance.

Table 2.6 shows the initial subsystems matrix that can be considered generic and common to vapor compression-based products.

System	Subsystems	
	ID	Name
Heating Ventilation Air Conditioning & Refrigeration	1	Human Machine Interface Subsystem
	2	Powertrain Subsystem
	3	Controls Subsystem
	4	Compression Subsystem
	5	Condensation Subsystem
	6	Expansion Subsystem
	7	Evaporation Subsystem
	8	Power Generation Subsystem
	9	Power Distribution Subsystem
	10	Heat Generation Subsystem
	11	Energy Recovery Subsystem
	12	Energy Storage Subsystem
	13	Health Management Subsystem
	14	Water Flow Subsystem
	15	Enclosure Subsystem
	16	Package Subsystem
	17	Documentation Subsystem

Table 2.6 Example of an HVAC & R Standard Subsystems

Once the main elements have been identified in this model, the next step is to perform a functional decomposition to get to the modules (still generic) that perform those functions or set of functions.

The identified subsystems apply to any Heating, Ventilation, Air Conditioning, and Refrigeration system. Whether it is a large commercial, medium commercial, residential, or transport system, it is to be noted that not all the subsystems, nor all the modules will be present in every application.

Considering some modules of the subsystems derived from the refrigerant cycle, we could have what is summarized in Fig. 2.10.

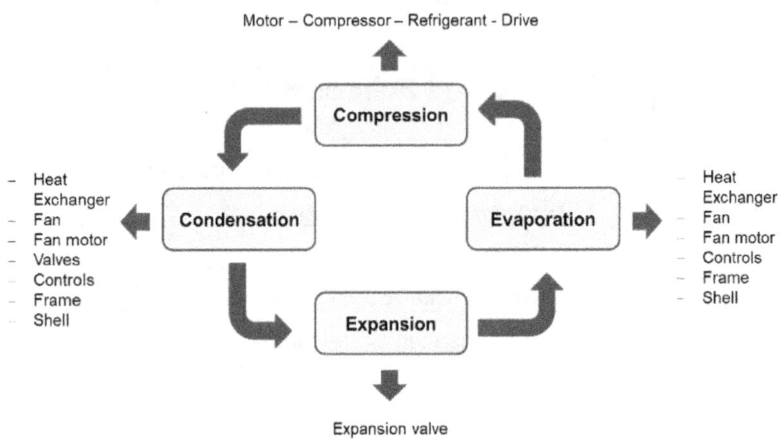

Fig. 2.10 Refrigerant Flow: Subsystems and Modules

Once the functional architecture is set, and all subsystems and modules are defined, the company is advised to create its product Portfolio Synoptic Matrix or PSM. Each product portfolio should have a PSM. Its strategic use is described in Chapter 4.

The PSM is one of the most important documents of any modular product management, driving tactical and strategic decisions. Table 2.7 represents the VCBA's example.

Table 2.7 is only illustrative and is not representative of any specific manufacturer's product. Its only purpose is to explain how existing products can be broken down from physical legacy to a functional architecture using the Operand Flow Analysis (OFA) method.

System	Subsystems		Modules		Comp.	Product Offerings			
	ID	Name	ID	Name		Large Commercial	Medium Commercial	Residential	Transport
Heating Ventilation Air Conditioning & Refrigeration	1	Human Machine Interface	1.1	Display	Major Components (not shown)	x	x	x	x
			1.2	HMI Supervisor (SW)		x	x	x	x
	2	Powertrain	2.1	Motor		x	x	x	x
			2.2	Engine		-	-	-	x
			2.3	Fuel		-	-	-	x
			2.4	Fuel Tank		-	-	-	x
			2.5	Drive		x	x	x	x
			2.6	Power Transmission		x	x	x	x
	3	Controls	3.1	Central Computing		x	x	x	x
			3.2	Control Panel		x	x	x	x
			3.3	Starter		x	x	x	x
			3.4	Unit Control Supervisor (SW)		x	x	x	x
			3.5	Group Control Supervisor (SW)		x	x	x	x
			3.6	Control Sensors		x	x	x	x
			3.7	Gateway		x	x	x	x
	4	Compression	4.1	Compressor		x	x	x	x
			4.2	Refrigerant		x	x	x	x
	5	Condensation	5.1	Heat Exchanger		x	x	x	x
			5.2	Fan		x	x	x	x
			5.3	Fan Motor		x	x	x	x
			5.4	Valves		x	x	x	x
			5.5	Frame		x	x	x	x
			5.6	Shell		x	x	x	x
	6	Expansion	6.1	Expansion Valve		x	x	x	x
			6.2	Expansion Valve Supervisor (SW)		x	x	x	x
	7	Evaporation	7.1	Heat Exchanger		x	x	x	x
			7.2	Fan		x	x	x	x
			7.3	Fan Motor		x	x	x	x
			7.4	Valves		x	x	x	x
			7.5	Frame		x	x	x	x
			7.6	Shell		x	x	x	x
			7.7	Economizer		x	x		x
	8	Power Generation	8.1	Generator		-	-	-	x
			8.2	Generator Prime Mover		-	-	-	x
			8.3	Generator Coupling		-	-	-	x
			8.4	Generator Switches		-	-	-	x
			8.5	Generator Control		-	-	-	x
			8.6	Generator Control Supervisor (SW)		-	-	-	x
			8.7	Ausiliary Power Generator		x	x	x	x
	9	Power Distribution	9.1	High Voltage Wiring		x	x	x	x
			9.2	Low Voltage Wiring		x	x	x	x
			9.3	Electrical Panel		x	x	x	x
			9.4	Switches		x	x	x	x
	10	Heat Generation	10.1	Electric Heater		-	x	x	-
			10.2	Electric Heater Control		-	x	x	-
			10.3	Electric Heater Supervisor (SW)		-	x	x	-
			10.4	Gas Heater		-	x	x	-
			10.5	Gas Heater Control		-	x	x	-
			10.6	Gas Heater Supervisor		-	x	x	-
	11	Energy Recovery	11.1	Thermal Energy Recovery		x	x	-	-
			11.2	Thermal Energy Recovery Supervisor (SW)		x	x	-	-
			11.3	Mechanical Energy Recovery		x	x	-	-
			11.4	Energy Recovery Supervisor (SW)		x	x	-	-
	12	Energy Storage	12.1	Electrical Energy Accumulator		-	-	-	x
			12.2	Mechanical Energy Accumulator		-	-	-	x
			12.3	Energy Level Supervisor (SW)		-	-	-	x
	13	Health Management	13.1	Health Sensors		x	x	x	x
			13.2	Edge Computing		x	x	x	x
			13.3	Gateway		x	x	x	x
			13.4	Network Supervisor (SW)		x	x	x	x
	14	Water Flow	14.1	Water Container		x	-	-	-
			14.2	Pump		x	-	-	-
			14.3	Pump Motor		x	-	-	-
			14.4	Water Pipes		x	-	-	-
			14.5	Fan		x	-	-	-
			14.6	Fan Motor		x	-	-	-
			14.7	Water Flow Control		x	-	-	-
			14.8	Valves		x	-	-	-
			14.9	Shell		x	-	-	-
	15	Enclosure	15.1	Base		x	x	o	x
			15.2	Walls		x	x	o	x
			15.3	Doors		x	x	o	x
			15.4	Roof		x	x	o	x
			15.5	Frame		x	x	o	x
			15.6	Noise Isolation		x	x	o	x
			15.7	Shell		x	x	o	x
	16	Package	16.1	Crate		x	x	x	x
			16.2	Lifting		x	x	x	x
	17	Documentation	17.1	Installation Manual		x	x	x	x
			17.2	Service Manual		x	x	x	x
			17.3	Spare parts Leaflet		x	x	x	x
			17.4	Packing List		x	x	x	x
			17.5	Shipping List		x	x	x	x
			17.6	Labels		x	x	x	x

Table 2.7 Represents the VCBA's Example

A.1.4 Function Analysis System Technique (FAST)

Charles W. Bytheway developed FAST in 1964. It was adopted for function analysis when dealing with Value Engineering. The method uses an intuitive logic to check the functions. The functions horizontally from left to right are in a How and Why relationship showing the method's graphical model. What is of interest in this book's context is how the functions are created and decomposed and their dependencies.

The generic model shown below completes the picture of the method:

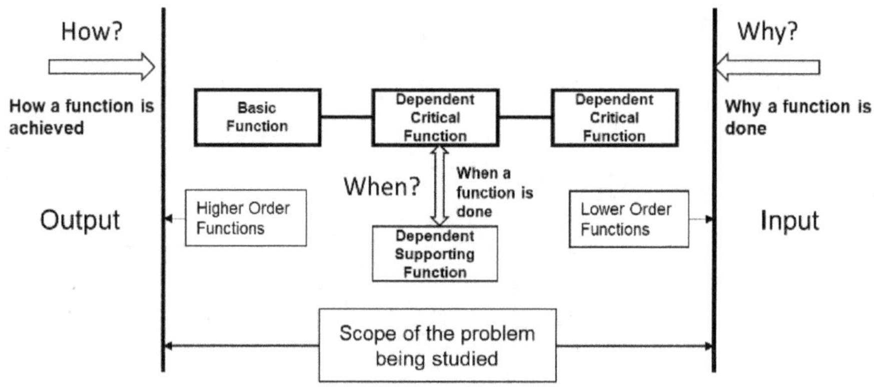

Fig. 2.11 FAST Generic Model

CHAPTER 3: SYSTEMS ARCHITECTURE

B1 Addendum to Section 3.4 Architecture Representations

B1.1 N² Diagram of a family of bikes

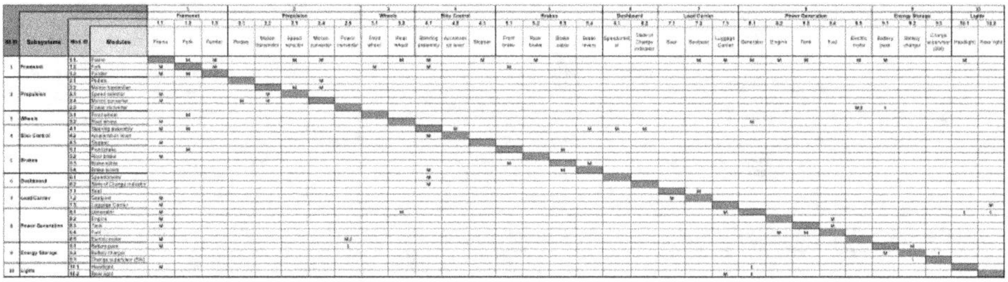

SS ID	Subsystems	Mod. ID	Modules	1			2				
				Frameset			Propulsion				
				1.1.	1.2	1.3	2.1	2.2	2.3	2.4	2.5
				Frame	Fork	Fender	Pedals	Motion transmitter	Speed selector	Motion converter	Power converter
1	Frameset	1.1.	Frame		M	M			M	M	
		1.2	Fork	M		M					
		1.3	Fender	M	M						
2	Propulsion	2.1	Pedals							M	
		2.2	Motion transmitter						M	M	
		2.3	Speed selector	M				M			
		2.4	Motion converter	M			M	M			
		2.5	Power converter								
3	Wheels	3.1	Front wheel		M						
		3.2	Rear wheel	M							
4	Bike Control	4.1	Steering assembly	M	M						
		4.2	Acceleration lever								
		4.3	Stopper	M							
5	Brakes	5.1	Front brake		M						
		5.2	Rear brake	M							
		5.3	Brake cable								
		5.4	Brake levers								
6	Dashboard	6.1	Speedometer								
		6.2	State of Charge indicator								
7	Load Carrier	7.1	Seat								
		7.2	Seatpost	M							
		7.3	Luggage Carrier	M							
8	Power Generation	8.1	Generator	M							
		8.2	Engine	M							
		8.3	Tank	M							
		8.4	Fuel								
		8.5	Electric motor	M							M,E
9	Energy Storage	9.1	Battery pack	M							E
		9.2	Battery charger								
		9.3	Charge supervisor (SW)								
10	Lights	10.1	Headlight	M							
		10.2	Rear light								

SS ID	Subsystems	Mod. ID	Modules	3.1 Front wheel (3 Wheels)	3.2 Rear wheel (3 Wheels)	4.1 Steering assembly (4 Bike Control)	4.2 Acceleration lever (4 Bike Control)	4.3 Stopper (4 Bike Control)	5.1 Front brake (5 Brakes)	5.2 Rear brake (5 Brakes)	5.3 Brake cable (5 Brakes)	5.4 Brake levers (5 Brakes)
1	Frameset	1.1.	Frame		M	M		M		M		
		1.2	Fork	M		M			M			
		1.3	Fender									
2	Propulsion	2.1	Pedals									
		2.2	Motion transmitter									
		2.3	Speed selector									
		2.4	Motion converter									
		2.5	Power converter									
3	Wheels	3.1	Front wheel									
		3.2	Rear wheel									
4	Bike Control	4.1	Steering assembly				M					M
		4.2	Acceleration lever			M						
		4.3	Stopper									
5	Brakes	5.1	Front brake								M	
		5.2	Rear brake									
		5.3	Brake cable						M			M
		5.4	Brake levers			M					M	
6	Dashboard	6.1	Speedometer			M						
		6.2	State of Charge indicator			M						
7	Load Carrier	7.1	Seat									
		7.2	Seatpost									
		7.3	Luggage Carrier									
8	Power Generation	8.1	Generator		M							
		8.2	Engine									
		8.3	Tank									
		8.4	Fuel									
		8.5	Electric motor									
9	Energy Storage	9.1	Battery pack									
		9.2	Battery charger									
		9.3	Charge supervisor (SW)									
10	Lights	10.1	Headlight									
		10.2	Rear light									

SS ID	Subsystems	Mod. ID	Modules	6.1 Speedometer (6 Dashboard)	6.2 State of Charge indicator (6 Dashboard)	7.1 Seat (7 Load Carrier)	7.2 Seatpost (7 Load Carrier)	7.3 Luggage Carrier (7 Load Carrier)	8.1 Generator (8 Power Generation)	8.2 Engine (8 Power Generation)	8.3 Tank (8 Power Generation)	8.4 Fuel (8 Power Generation)	8.5 Electric motor (8 Power Generation)
1	Frameset	1.1.	Frame				M	M	M	M	M		M
		1.2	Fork										
		1.3	Fender										
2	Propulsion	2.1	Pedals										
		2.2	Motion transmitter										
		2.3	Speed selector										
		2.4	Motion converter										
		2.5	Power converter										M,E
3	Wheels	3.1	Front wheel										
		3.2	Rear wheel						M				
4	Bike Control	4.1	Steering assembly	M	M								
		4.2	Acceleration lever										
		4.3	Stopper										
5	Brakes	5.1	Front brake										
		5.2	Rear brake										
		5.3	Brake cable										
		5.4	Brake levers										
6	Dashboard	6.1	Speedometer										
		6.2	State of Charge indicator										
7	Load Carrier	7.1	Seat				M						
		7.2	Seatpost			M							
		7.3	Luggage Carrier										
8	Power Generation	8.1	Generator					M					
		8.2	Engine									M	
		8.3	Tank									M	
		8.4	Fuel							M	M		
		8.5	Electric motor										
9	Energy Storage	9.1	Battery pack										
		9.2	Battery charger										
		9.3	Charge supervisor (SW)										
10	Lights	10.1	Headlight						E				
		10.2	Rear light					M	E				

SS ID	Subsystems	Mod. ID	Modules	9			10	
				Energy Storage			Lights	
				9.1	9.2	9.3	10.1	10.2
				Battery pack	Battery charger	Charge supervisor (SW)	Headlight	Rear light
1	Frameset	1.1.	Frame	M			M	
		1.2	Fork					
		1.3	Fender					
2	Propulsion	2.1	Pedals					
		2.2	Motion transmitter					
		2.3	Speed selector					
		2.4	Motion converter					
		2.5	Power converter	E				
3	Wheels	3.1	Front wheel					
		3.2	Rear wheel					
4	Bike Control	4.1	Steering assembly					
		4.2	Acceleration lever					
		4.3	Stopper					
5	Brakes	5.1	Front brake					
		5.2	Rear brake					
		5.3	Brake cable					
		5.4	Brake levers					
6	Dashboard	6.1	Speedometer					
		6.2	State of Charge indicator					
7	Load Carrier	7.1	Seat					
		7.2	Seatpost					
		7.3	Luggage Carrier					M
8	Power Generation	8.1	Generator				E	E
		8.2	Engine					
		8.3	Tank					
		8.4	Fuel					
		8.5	Electric motor					
9	Energy Storage	9.1	Battery pack		M			
		9.2	Battery charger	M		I		
		9.3	Charge supervisor (SW)		I			
10	Lights	10.1	Headlight					
		10.2	Rear light					

CHAPTER 5: SYSTEMS DESIGN

C1 System Requirements Key Sections

Section 1: Approvals and Version Control

- Document Approvals and Version Control: The system requirements are essential and technical by nature; any misunderstanding or failure to keep them current could result in a lousy design and offering with disastrous financial consequences. Therefore, once completed, they should be approved by the functional leaders of engineering, marketing, manufacturing, and supply chain. Any further changes should be version-controlled and approved by the initial functions or their representatives. The validation of the final system should be done with the most up to date document.

Section 2: Introduction

- Definitions and Acronyms: Every company has its own language expressed through special terms. Beyond such internal language, engineers use terminology that they often make up or draw from various industry standards and regulations. The system requirements document carries the burden to fully describe the system that will be built and supplied to customers. Therefore, a successful design starts from a good document that is unambiguous to those who will use it to design, build, assemble, and eventually install the system. This section covers all the acronyms used throughout the document, and the definition of the terms used may not be evident to all those involved.
- System Description: This section describes the mission of the system, with a summary of its main functions and features. It should be a high-level description that also captures the market segments addressed and any competitor product already being offered.

Section 3: Codes, Regulations, and Standards

- Codes and regulations: Products or, in general, systems designed to address specific market needs are often required to meet certain countries' regulations. In this section, the reference to such rules and country codes should be provided to allow the design and production team to comply. Some products even require approval of sales in a country only after successfully passing an inspection or acceptance/witness tests by inspection officials. The documents may not be included to avoid a voluminous and impractical system requirements document, but their references need to be summarized for compliance.

- Standards: Same as in the regulations section, the references to industry standards should be summarized in this section. Any certifying entity's specifications should also be referenced for compliance. Satisfying them is a condition for the system's release for public use.

Section 4: Functional Requirements

Referring to Fig. 5.4 above, this is the section where the primary functions of the system are specified. It is recommended that every business invests in performing a thorough functional analysis of its products to establish a functional architecture of its system offerings. (Refer to chapters 2, 3, and 4 above). The functional analysis should be done outside of any specific design in order to avoid the mistake of having it dictated solely by an existing and known physical architecture. It must be functional by nature, and the resulting functional architecture should set the subsystems and modules. The architecture document is a precious asset for the company as it establishes how products are designed, manufactured, and distributed.

The functional requirements are expressed as performance and operational requirements. Performance requirements qualify the functions by expressing how well they must be performed to meet the stakeholders' expectations. As operational requirements, they specify how the system should operate.

A brief description of each of the elements of the system's functional requirements would help better understand them.

- Performance Requirements: Each of the system-level functions should be expressed with its metrics, meaning how well the system performs to fulfill its mission. An example would make it easier to understand. For an elevator, the following performance requirements are often used:
  - The car shall move at speed equal to 1.6 m/s.
  - The system shall provide ride comfort. The maximum vibration

levels shall not exceed 8 mg horizontal (side to side and front to back) and 8 mg vertical. The noise in the car expressed as sound pressure shall not exceed 50 dBA.

- Operational requirements: As stated in Fig. 5.4 and the section above, functional requirements can also be expressed as operational requirements that would supplement the performance requirements. Continuing with the example of the elevator, some operational requirements are:
    - The elevator shall acquire the passengers' service request. (Even though the current solution is through pressing a button, the requirement should be generic enough to allow the exploration of other solutions during the conceptual design phase). It shall also acknowledge the request by providing feedback within 1.5 seconds. (The current answer is a light that comes on around or inside the button when pushed. In the industry, it is called a telltale light).
    - The elevator control shall operate with a single push button, in which case the car shall go to the landing where the request was made, to load the passenger or operate in a collective mode and collect all the requests and stop the car at all the floors with requests.
    - The system shall perform an advanced door opening (ADO) 0.5 – 1.0 second before stopping, to reduce the flight time (the travel time from floor to floor).
    - The system shall display the car's position and or direction both inside the car to the traveling passengers and on the floor to the waiting passengers.
    - The system shall announce the arrival of the car at a floor 4.0 seconds before stopping and opening the doors.

If instead of an elevator, an air conditioner was considered, an example of operational requirement would be:

- If the outside temperature is less than the indoor temperature (by x°C degrees Celsius), the system shall intake, filter, and circulate outside air into the environment to cool to save energy.

Those are a few examples of operational requirements based on system functions. All the metrics used are, for example only.

Section 5: Non-Functional Requirements

As equally important as the functional requirements, the non-functional requirements need to be met to deliver a successful product, a service, or an organization. They generally describe the system and its attributes and constraints.

The following paragraphs cover a few items that could be part of the non-functional components of the requirements document.

- **Performance requirements**: Performance requirements include a specific set of metrics with defined values that translate how well the system attributes meet the stakeholders' expectations. As expressed earlier, performance requirements can be both functional and non-functional. In the case of an elevator, the reliability is an important requirement. Its availability is an important differentiator and can drive a competitive advantage for service contracts. No user or building owner wants an elevator that is not responsive when needed, especially in hotels, hospitals, etc. The maximum rise of the system or the maximum number of floors its passenger cabins can serve is another example of performance requirement and yet non-functional.
- **Operational requirements**: This is the section that describes how the system will be used, what it needs, to operate as intended, and its interaction with its operating environment and with other systems.
    - Operating environment: it describes temperature range, relative humidity, altitude in which the system should work in normal mode. If motors are involved, for example, all additional related requirements should be specified; (power requirements: voltages, frequencies, etc. or how they could be de-rated due to altitude). Addressing companies' sustainability goals, the impact of the product on the environment should also be specified.
    - Operating modes: this section describes the operation of the system. The typical questions to answer are:
    - How shall the system be set up? What are the startup steps and parameters? What are the standard operating parameters to meet the expected performance levels? What are the failure modes or abnormal modes of operation? Graceful degradation or sudden shutdown?
    - What should the system's behavior be in case of catastrophic events due to unusual inputs (earthquake, fire, cyber-attack)?

- **Logistics requirements**: A well-designed system and assembled in the factory can be damaged during its transportation to its operating site under certain circumstances. Therefore, logistics requirements should not be overlooked. Improvising, after the fact, is not a good idea. In today's environment, logistics is critical to the success of any business. It should thoroughly be specified and analyzed. Often, packing, loading, shipping, and transportation are not conceived by the system's architects or designers but left to people not involved in the system's design's rigorous phases. This lack of inclusion at the design stage can compromise significant investments at the end of the process. A product's behavior at its operating site should be the same as during its validation or final acceptance testing, indicating that no damage occurred during handling and transportation. However, if proper care is disregarded, a whole troubleshooting phase could start at the customer site, leading to customer dissatisfaction or financial loss. Packaging design, be it a wooden crate or cardboard, is as important as the product itself and considers the product's exposure to the elements, the road conditions, vibration, for example, and any other business-specific conditions.
- **Size (envelope) and footprint**: In this section, the natural constraints of the system, size-wise, are to be specified. For example, any system designed for installation in a house or apartment will create a whole set of issues if its package does not have at least one dimension that can pass through the doors. As I am often fascinated by what is happening around me, I stopped for a few minutes to see how a crew of movers trying to lift a piano to the eighth floor of an apartment in Paris (France) would solve the problem. The crane operator was very skillful, but both the balcony and the door were too small for the giant piano. After waiting for a while, with a hanging package and people strategizing around, I left. When I walked by hours later, they were gone, and I assumed they found the way to get the system in or simply gave up. The takeaway here was that a system designed with some constraints in mind would be differentiated enough to present a competitive edge.
- **Installability, maintainability, and serviceability**: This section shall specify the installation, maintenance, and service requirements. The designers will then comply by making the right trade-offs.
- For installability, the desired time to install the system, if it is not a plug

and play, is critical to minimize labor cost. Once factory labor cost is set in a controllable environment, installation cost becomes a competitive differentiator, minimizing the system's lead-time from customer order to its operation at the customer site. In a modular product environment, well-defined interface requirements provide a means of meeting this need.

Once a system is professionally installed and put in service, its maintainability and serviceability can be a source of competitive advantage. With proper health monitoring, unplanned downtime can be minimized and even eliminated. In case of an eventual breakdown, the established requirements of time to diagnose and repair will allow a quick intervention to put the system back into service. How the system should be maintained and serviced to eliminate unplanned downtime requires a careful definition during the requirements phase.

- **Aesthetics, Finishes, colors:** No matter what a system is designed to do, aesthetics always make a difference in the eyes of the buyer. Even systems designed to be used as subsystems of a bigger system still need to look aesthetic. Although beauty is subjective and not often considered for human-made objects other than artwork, a system, or merely any device with round edges, smooth skin, manicured details will first get everyone's attention than others without such features. The choice of finish colors should also be well thought with expert advice depending on the fields of application of the system. Colors should be inviting and inspires cleanness and appear non-threatening. Medical applications of any system should have appropriate colors. A factory floor, a school, a restaurant, etc., should have colors carefully chosen to give the right character to the place.
- **Human Interface and Ergonomics**: This section specifies the requirements for human interaction with the system. Any stress on the operator should be minimized. Ergonomic features and simple communication with the system through displays and any other means should be considered to allow the designers to deliver a differentiated product.
- **Reliability**: In my public presentations and talks, I often explain the difference between quality and reliability by assuming that, typically, quality should be when a system does what it is designed to do. In

other words, a system that fully meets its requirements is a system of good quality. Unfortunately, customers do not often know what the requirements of the system were. They only have their expectations, and a system of a good quality quickly becomes a system that does what they expected it to do. The ability of the system architects to translate such customer expectations, as unmet needs, into requirements drives the success or failure of a product, a process, a service, or even an organization. This ability is a combination of art and science and comes with excellent knowledge of the market, the buying behaviors, and future scenarios planning. Once the quality is understood in those terms, a reliable system can be defined generically as a system that does what it is expected to do over a long period of time without any unwanted interruption. So, reliability is quality over time, with quality broadly defined from the buyer or user's view.

- **Cost**: it is another non-functional requirement. A lot has been written in the literature about cost requirements. Different from price requirements generally linked to selling strategic choices of a company, the cost is a fact that must be well understood and designed to meet. It is also an object of technical trade-offs; but no matter how it is dealt with, it is important to consider all the different types of cost:
    - The product/system factory cost: it is mostly made of the material and factory labor cost.
    - The cost at delivery (also called landed cost) should include the factory cost defined above with its packing, loading, shipping, and transportation cost. By setting a target for the landed cost, the designers pay attention to the shipping and transportation costs that are quickly impacted by the size of the packages. Details such as container size, truck size (avoid flatbed trucks), or even door sizes are important to be considered to minimize loading and unloading, shipping, and transportation costs.
    - The total lifecycle cost: it includes the purchase price (which may or may not include installation cost), the operating cost, maintenance or service cost, any upgrade cost, and the remaining residual value at the end of ownership or its first use. It is important to highlight that the sum of the purchase price, operating cost, and any eventual replacement or upgrade cost is also defined as the total cost of owner-

ship (TCO), which becomes relevant when comparing systems to be operated for a long time.

Systems sold that require service over time should be well defined at the requirements phase. Customers are now paying more attention to the total cost of ownership. Savvy customers may no longer consider purchasing an initially inexpensive system that requires a high operating cost. Service terms are evolving beyond servicing oil and grease with repair maintenance when the system stops, to become the selling of uptime operation of the system. It is viewed as insurance where the customers contract a service agreement to have peace of mind and have the system operate every time it is expected to do so.

- **Safety**: Depending on the range of the product's users, public safety, operator safety, and equipment safety requirements should be defined in terms of potential operator hazards during manufacturing and assembly, installation, and service. Potential risks for users and, in general, the public should be identified and addressed.
- End of Use: The world has evolved to the point where products should be developed with the end of life or more appropriately "end of use" in mind and well defined. A product designed with zero landfill or total recycle-ability as a requirement will need some specific choices of material during its conception.

Section 6: Subsystem and Module Requirements

The system-level requirements defined in the previous sections now need to be allocated to the subsystems and from the subsystems to the modules following the standard system functional structure. In this case, the word standard assumes prime importance. The system's architecture definition from functional analysis and grouping should have happened outside of the current design. Specific development should focus on defining the physical architecture based on the choice of the modules' physical instances, their variants, and interfaces. It is the phase known as synthesis in systems engineering.

One of the fascinating exercises in requirements engineering is their allocation to the lower level system elements (subsystems, modules, and even to components). A top-down approach reviews each system-level requirement and maps it to all the subsystems that contribute to its achievement. When several subsystems concurrently contribute to fulfilling a requirement, all of them are tagged with it. The next phase of the exercise is to define the metrics relative to each subsystem. In general, a higher-level requirement is fully allocated when all its lower-level contributors have been identified, and their requirements fully defined for each one of them with the right metrics when applicable.

Fig. 5.5 illustrates one allocation scenario where each type of requirement is decomposed and mapped to the subsystems to drive its fulfillment. Not all system-level requirements impact every subsystem.

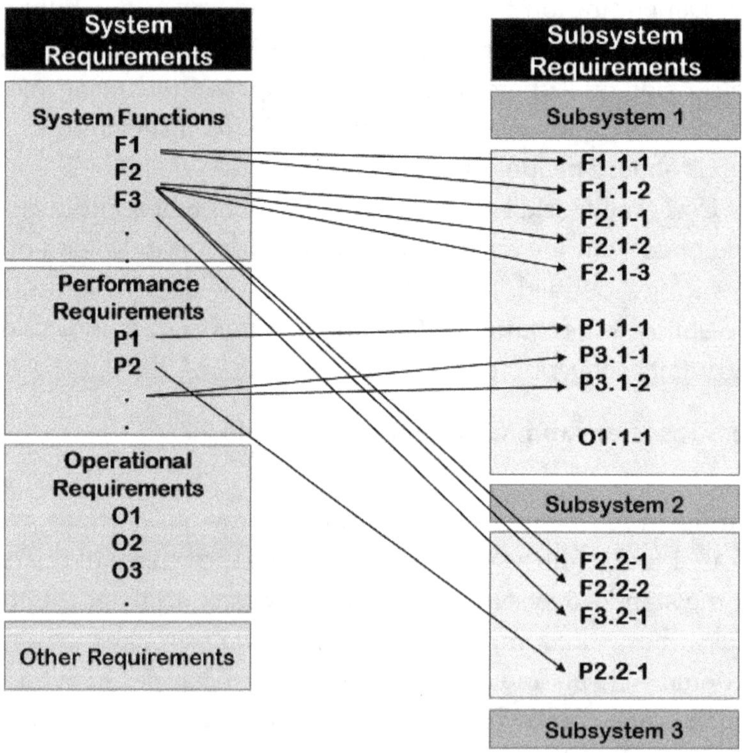

Fig. 5.5 System requirements allocation to subsystems

As shown in the generic allocation example of Fig. 5.5 above, the system's functions that determine the functional requirements are F1, F2, F3, etc. They drive the performance requirements P1 and P2 and the operational requirements O1, O2, and O3. F1 is allocated to subsystem 1 and decomposed into two sub-functions F1.1-1 and F1.1-2.

F1.1-1 representation should be read as system functional requirement 1 allocated to subsystem1 (as subsystem functional requirement 1). Then F1.1-2 as system functional requirement 1 allocated to subsystem 1 (as subsystem functional requirement 2).

F2 is allocated to subsystem 1 and subsystem2.

A system example to illustrate the allocation mechanism concept. The system to be considered is the simple house below:

Fig. 5.6 A simple house

The simple house has the following subsystems and modules:

System = Simple House						
Subsystems		Modules				
1	Roof	1.1 shingles	1.2 structure	1.3 Gutters		
2	Walls	2.1 Front Wall (South)	2.2 East Wall (Side)	2.3 West Wall (Side)	2.4 Back Wall (North)	
3	Doors	3.1 Front Door	3.2 Back Door			
4	Windows	4.1 Front Window	4.2 East Window	4.3 West Window		
5	Floor	5.1 Carpet	5.2 Structure			

Fig. 5.6 B Simple House Elements

The next figure illustrates the elements at different levels:

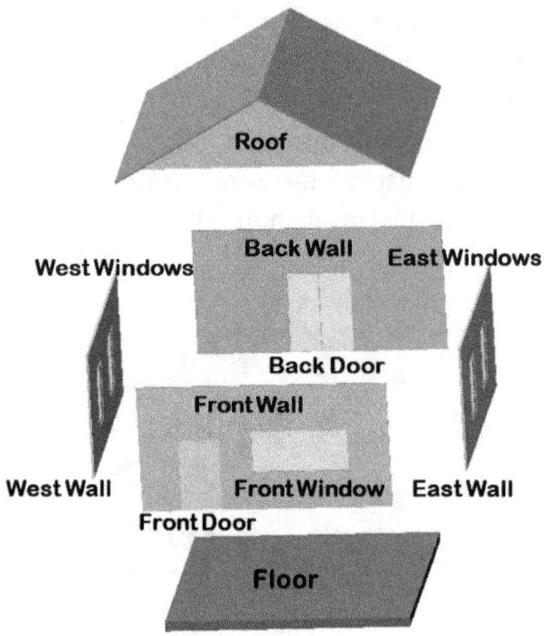

Fig. 5.6 C Simple House Elements

Two requirements on the house are:

1. The house shall support one-meter deep of snow load
2. Entry and exit to and from the house shall be possible with keys supplied to the tenant

The two requirements highlight two essential functions:

a. Support load (from snow)
b. Open and Close door

The allocation to the subsystems is illustrated in Fig. 5.7. The allocation exercise continues throughout all the system requirements with the right labeling. Once completed, each subsystem has its own set of requirements, allocated to the modules that compose them. In Fig. 5.7, subsystem 4 and 5 are

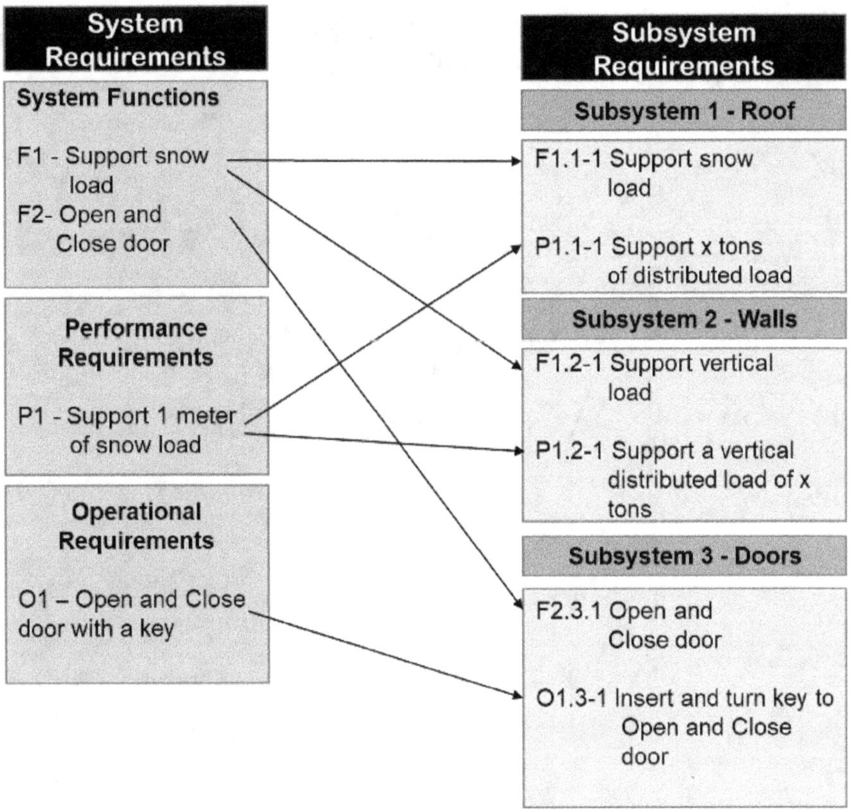

Fig. 5.7 Simple house system functions and requirements allocated to its subsystems

not represented. Using the same simple house example, the two requirements are allocated from the subsystems to their modules, as shown in Fig. 5.8.

As the exercise continues, subsystems 3, 4, and 5 will be added, and their requirements allocated to the modules. In general, the stakeholder or marketing requirements would be allocated to the system and the subsystems. The allocation continues to modules and sometimes even to some significant components, especially complex and mostly not off the shelf.

After the allocation, most, if not all, module requirements should be traceable to the subsystems requirements, and ultimately to the system requirements. The impact of traceability should not be underestimated. Requirements reviewers are responsible for assuring that all higher-level

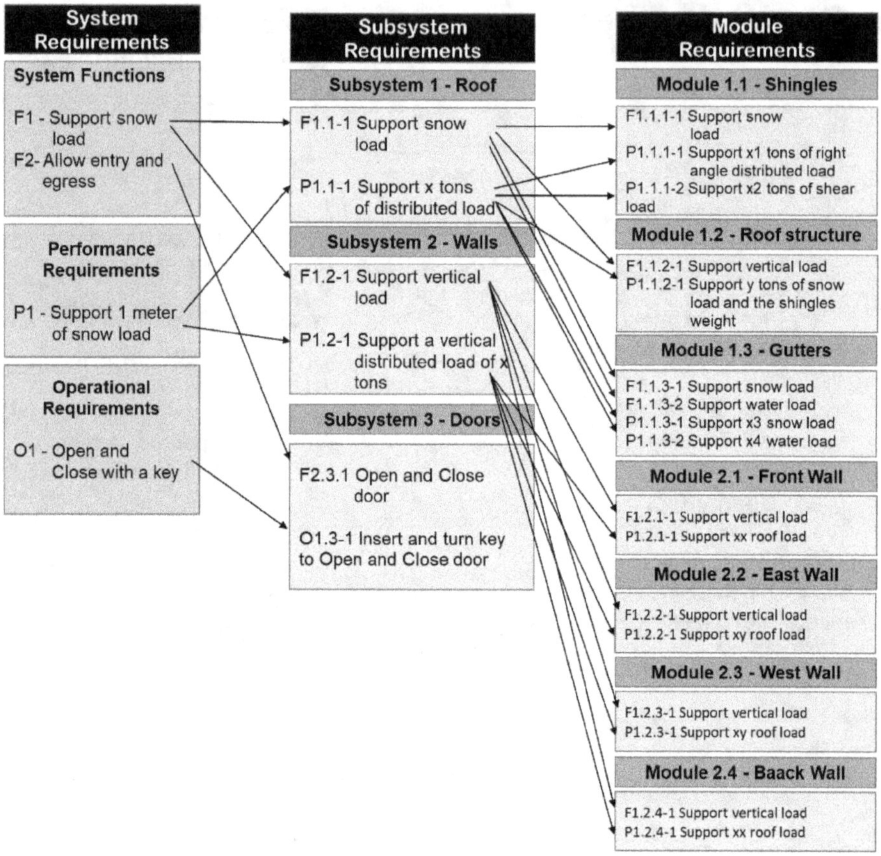

Fig. 5.8 Simple house subsystem requirements allocated to its modules

requirements are addressed at the appropriate lower entities' level and that all lower-level requirements trace back to a Stakeholder need. The notion is summarized in Fig. 5.9.

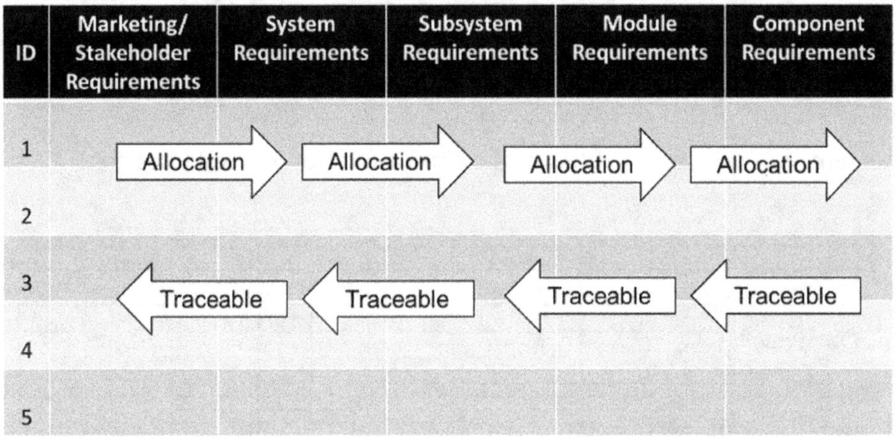

Fig. 5.9 Requirements table with Allocation and Traceability directions

CHAPTER 7: MANUFACTURING & SUPPLY CHAIN

D1 Addendum to Section 7.2.2.1 System Requirements

This template can be adopted and be used for Reference Units cost.

SS ID	Subsystems	Mod. ID	Modules	Modules Instances/ Variants #	Components	Unit Cost	Quantity	Module Cost	Subsystem Cost	Comments
1	Frameset	1.1.	Frame							
		1.2	Fork							
		1.3	Fender							
2	Propulsion	2.1	Pedals							
		2.2	Motion transmission							
		2.3	Speed selector							
		2.4	Motion converter							
		2.5	Power converter							
3	Wheels	3.1	Front wheel							
		3.2	Rear wheel							
4	Bike Control	4.1	Steering assembly							
		4.2	Acceleration lever							
		4.3	Stopper							
5	Brakes Assembly	5.1	Front brake							
		5.2	Rear brake							
		5.3	Brake cable							
		5.4	Brake levers							
6	Dashboard	6.1	Speedometer							
		6.2	Charge indicator							
7	Load Carrier	7.1	Seat							
		7.2	Seatpost							
		7.3	Luggage Carrier							
8	Power Generation	8.1	Generator							
		8.2	Engine							
		8.3	Tank							
		8.4	Fuel							
		8.5	Electric motor							
9	Energy Storage	9.1	Battery pack							
		9.2	Charge supervisor (SW)							
10	Lights	10.1	Headlight							
		10.2	Rear light							
			Total System Cost						SUM(Rows1-10)	

Table 7.1 Bike reference material cost model template

SS ID	Subsystems	Mod. ID	Modules	Components	Module Material Cost	Module Assembly Labor Cost	Total Module Cost	Total Subsystem Cost	Comments
1	Frameset	1.1.	Frame	Frame, Rear Shock absorber					
		1.2	Fork	Fork, Bearings, Nuts, Front shock absorber					
		1.3	Fender	Front fender, Rear fender					
2	Propulsion	2.1	Pedals	Right pedal, Left pedal					
		2.2	Motion transmission	Chainrings, Crankarms, Spider, Chain					
		2.3	Speed selector	Cassette, Front derailleur, Rear derailleur					
		2.4	Motion converter	Crankset					
		2.5	Power converter	Drive					
3	Wheels	3.1	Front wheel	Rim, Hub, Spokes, Bearings, Tube, Tire					
		3.2	Rear wheel	Rim, hub, spokes, bearings, tube, tire					
4	Bike Control	4.1	Steering assembly	Handlebar, Headset, Stem, Grips					
		4.2	Acceleration lever	Acceleration Lever/Shifter					
		4.3	Stopper	Stopper					
5	Brakes Assembly	5.1	Front brake	Disc, Caliper, Pad					
		5.2	Rear brake	Disc, Caliper, Pad					
		5.3	Brake cable	Front cable, Rear Cable					
		5.4	Brake levers	Front lever, Rear Lever					
6	Dashboard	6.1	Speedometer	Speedometer, Wire, Wheel hub connector					
		6.2	Charge indicator	Display, Wire, Battery clamp					
7	Load Carrier	7.1	Seat	Saddle, Saddle clamp					
		7.2	Seatpost	Seatpost, Seat clamp					
		7.3	Luggage Carrier	Luggage Carrier					
8	Power Generation	8.1	Generator	Generator, Generator clamp					
		8.2	Engine	Engine, Clutch, Exhaust pipe,					
		8.3	Tank	Container, Filter, Cap, Hose, Tank clamps					
		8.4	Fuel	Fuel					
		8.5	Electric motor						
9	Energy Storage	9.1	Battery pack	Battery, Battery clamp, Charge electronics					
		9.2	Charge supervisor (SW)	State of charge software					
10	Lights	10.1	Headlight	Headlight, Headlight clamp, Wire					
		10.2	Rear light	Rear light, Rear light clamp, Wire					
			Total System Cost					SUM(Rows1-10)	

Table 7.2 Total Module and Subsystem Cost (including assembly cost)

A FEW SYSTEMS ARCHITECTURE PRINCIPLES

My professor of Systems Architecture at Massachusetts Institute of Technology (MIT), E. Crowley, encouraged each of his students to continually think of Systems Architecture principles. The Merriam-Webster dictionary defines a principle as "a basic truth or theory," and I align well with such a definition. In fact, in Latin, a principle which is principium means "beginning or starting point." I found the exercise remarkably interesting and kept my list, which I used in some of the book's concepts. For reference, I have picked ten principles that are relevant to the book from the list, as shown below:

1. Every natural or engineered system performs a function or set of functions.
2. Every system is the subsystem of a larger system.
3. Every system, natural or engineered, continuously evolves.
4. Every system needs some form of energy to function.
5. Every engineered system meets both functional and non-functional requirements.
6. No single system lives or operates in isolation.
7. Humans live in systems of systems.
8. Every person crosses multiple system boundaries countless times every day of his or her life.
9. Every system of systems expands towards the universe (infinitely big)
10. Continuing to break a system into its components tends to drive toward nanoparticles and beyond (infinitely small).

Glossary and Acronyms

A

Architecture: In general, the architecture of a system can simply be defined as the integrated structure of its elements, their behaviors, and their relationships. In a more formal and widely agreed-upon statement, ISO 42010 defines architecture as the fundamental concepts or properties of a system in its environment embodied in its elements, relationships, and in the principles of its design and evolution

ATO: Assemble to Order

B

B2B: Abbreviation of Business to Business

B2C: Abbreviation of Business to Consumer

BOM: Bill of Materials

C

CAD: Computer Aided Design

CCB: Change Control Board

CR: Change Request

D

DDI: Demand-Driven Innovation. Innovation to satisfy a well-defined demand (usually from B2B)

DFI: Design for Installation

E

EPBA: Existing Product Breakdown Analysis – is a functional analysis method based on the decomposition of existing product elements to extract their functions that are then regrouped at a level above where the previous design decisions were made.

ETO: Engineered to Order

F

Function: Simply put, a function is "what a device or, in general, an entity does; it is the task or series of tasks that the entity in consideration needs to perform to achieve the desired outcome or intended objective." Hubka and Eder [1] define a function as "a property of the technical system which describes its ability to fulfill a purpose, namely, to convert an input measure into a required output measure under precisely given conditions."

FRU: Field Replaceable Unit

G

H

Hub module: A module that is connected to many other modules.

I

ICD: Interface Control Document

IDR: Independent Design Review

Instance: Category of module

Interface: NASA Reference Publication 1370 of January 1997 defines an interface as "A design feature of a piece of equipment that affects the design feature of another piece of equipment

IPD: Integrated Product Development (is usually applied to teams with members from product management, engineering, manufacturing, supply chain, field installation, service, finance, HR, etc.)

ISKU: Incremental Solution to Known Unmet (needs)

ISUU: Incremental Solution to Unknown Unmet (needs)

J

K

L

LRU: Lowest Replaceable Unit

M

Module: In this study, a module is defined as an independent assembly designed to perform a set of functions and deliver an expected output.

MPA: Modular Product Architecture

MRD: Market Requirements Document

MSP: Module Supply Partner

MTO: Make to Order

MTS: Make to Stock

N

NPD: New Product Development

N-Square Diagram: A diagram made of a system's modules on each side to express their relationships. The number of rows equals the number of columns in an NxN representation.

O

ODI: Opportunity Driven Innovation

OEM: Original Equipment Manufacturer

OSR: Operational Scenarios Review (Method) – is a method based on reviewing the succession or sequence of actions that any person interacting with a system would take.

P

Platform: A product platform is a defined set of subsystems or modules arranged to be re-used across the offerings of a portfolio.

PSM: Portfolio Synoptic Matrix

Q

R

RPCC: Reference Product Configuration Cost

RSKU: Radical Solution to Known Unmet (needs)

RSUU: Radical Solution to Unknown Unmet (needs)

S

SOI: System of Interest

SRD: Stakeholders' Requirements Document

SSP: Subsystem Supply Partner

Subsystem: A subsystem is a subordinate group of elements of a system. In a system of systems, a subsystem could be itself a system

SyRD System Requirements Document

System: Generally, the term System can be applied to any entity made of integrated elements forming a unified whole with a purpose not achievable by any of its constituents

T

TBM: Time Based Maintenance

TCO: Total Cost of Ownership

TPM: Total Productive Maintenance

U

V

VAR: Value Added Reseller

Variant: Module instance with specific distinguishable characteristics

VOC: Voice of the customer

V&V: Verification and Validation

W

WinnoveX: Winning Innovation Excellence

X

Y

Z

References

1. *Theory of Technical Systems*, Vladimir Hubka & W. Ernst Eder, Editions Springer-Verlag 1988 p. 72
2. *Encyclopedia Britannica*
3. *The Principles of Design*, Nam P. Suh, Oxford University Press 1990
4. ISO-42010
5. USPTO, Patent US4665492A, Computer Automated Manufacturing process
6. NASA Reference Publication 1370, January 1997
7. *Factory Physics*, 2nd Edition, Irwin McGraw-Hill 2001, p.1
8. *Emotional Design*, Norman, Donald A., Basic Books 2004, p.11

Selected Bibliography

1. Blanchard, S. Benjamin; Fabrycky, Wolter J., *Systems Engineering and Analysis*, 5th Edition, Prentice-Hall International 2011
2. Norman, Donald A., *Emotional Design*, Basic Books 2004
3. Eppinger, D. Steven, and Browning, R. Tyson, *Design Structure Matrix Methods and Applications*, The MIT Press Cambridge, Massachusetts 2012
4. Hubka, Vladimir and Eder; W. Ernst, *Theory of Technical Systems*, 2nd Edition, Springer-Verlag Berlin, Heidelberg 1984
5. INCOSE, *Systems Engineering Handbook*, 4th Edition, Wiley, Hoboken, New Jersey 2015
6. Kossiakoff, Alexander; Sweet, N. William; Seymour, J. Samuel and Biemer, M. Steven, *Systems Engineering: Principles and Practice*, 2nd Edition Wiley Hoboken, New Jersey 2011
7. Nelson, Tim, and McGee, Jim, *Think Inside the Box*, WCG Press, Winnetka, IL 2013
8. Project Management Institute (PMI), *Project Management Body of Knowledge (PMBOK) Guide*, 5th Edition PMI 2013
9. Robertson, Suzanne, and Robertson, James, *Mastering the Requirements Process*, 2nd Edition, Addison-Wiley 2006
10. Simpson, W. Timothy; Jiao, Jianxin (Roger); Siddique, Zahed and Holtta-Otto, Katja, *Advances in Product Family and Product Platform Design*, Springer New-York, NY 2014
11. Steven, Richard; Brook, Peter; Jackson, Ken, and Arnold, Stuart, *Systems Engineering: Coping with Complexity*, Prentice Hall Europe 1998
12. Suh, Nam P., *The Principles of Design*, Oxford University Press 1990
13. Sweeney, John, *Innovation: At the Speed of Laughter*, Aerialist Press, Minneapolis, MN 2004
14. Watson, S. Charles, *System Analysis, Design, and Development: Concepts, Principles, and Practices*, Wiley Hoboken, New Jersey 2005
15. Westrick, Rob, and Cooper, Chris, *Winning by Design*, 2012

About the Author

Leandre Adifon is the Founder and CEO of Pyramid Base Technologies, a company that makes Earth-friendly high tech affordable to those who live at the base of the social, economic pyramid and help other companies innovate.

He has spent his career bringing new products to life and has authored and co-authored more than 40 patents in the companies he has worked for. Born in the Republic of Benin (West Africa), he grew up in Africa and Europe. Now he lives in the United States of America as a US citizen.

Leandre has works for large multi-national companies developing his natural gifts to become a creative, adaptive and results-oriented executive with extensive engineering and operations experience in manufacturing and service industries.

He is a recognized and sought leader for Business and Innovation Vision setting with a record of building, leading and delivering world-class, global technology platforms, products and teams combined with a well-documented history of leveraging the technological excellence of an organization as a driver of business outcomes.

Formerly Leandre was Vice President at Ingersoll Rand PLC, the diversified industrial company that made commercial, residential and transport HVACR with other industrial equipment, in the role of Enterprise Systems Engineering and Advanced Technology leader.

Leandre came to Ingersoll Rand after 20 years at United Technologies, where he was Vice President of Otis Worldwide Engineering and a member of the Otis Executive Committee, preceded by roles of increasing responsibility.

Prior to joining United Technologies, Leandre was consultant in Robotics, machine tools and thermal engineering in Europe.

He holds advanced degrees in mechanical engineering from Polytechnic of Milan in Italy and in management and finance technology from Rensselaer Polytechnic Institute (RPI) in the US. He also has an MBA from RPI and a certificate in systems design and management from Massachusetts Institute of Technology (MIT).

Leandre speaks several languages, including French, Italian, Spanish and some German.

Index

Symbols

A

B

C

D

E

N

O

P

R

S

T

U

www.ingramcontent.com/pod-product-compliance
Lightning Source LLC
LaVergne TN
LVHW080020210926
848481LV00011B/3070